Suisun Marsh

The publisher gratefully acknowledges the generous contribution
to this book provided by the Water Resources Center,
University of California, Davis.

Suisun Marsh

Ecological History and Possible Futures

———

Edited by

Peter B. Moyle, Amber D. Manfree,
and Peggy L. Fiedler

UNIVERSITY OF CALIFORNIA PRESS

Berkeley   Los Angeles   London

University of California Press, one of the most distinguished university presses in the United States, enriches lives around the world by advancing scholarship in the humanities, social sciences, and natural sciences. Its activities are supported by the UC Press Foundation and by philanthropic contributions from individuals and institutions. For more information, visit www.ucpress.edu.

University of California Press
Berkeley and Los Angeles, California

University of California Press, Ltd.
London, England

Library of Congress Cataloging-in-Publication Data

Suisun marsh : ecological history and possible futures / edited by Peter B. Moyle, Amber D. Manfree, Peggy L. Fiedler.
    pages   cm
  Includes bibliographical references and index.
  ISBN 978-0-520-27608-6 (paperback)
  ISBN 978-0-520-95732-9 (e-book)
    1. Suisun Marsh (Calif.)—History.   2. Marsh ecology—California—Suisun Marsh.   3. Salinity—California—Suisun Marsh.   4. Brackish water ecology—California—Suisun Marsh.   5. Water quality—California—Suisun Marsh.   I. Moyle, Peter B.
  QH105.C2S874   2014
  577.6809794'6—dc23   2013048757

Manufactured in the United States of America

23   22   21   20   19   18   17   16   15   14
10   9   8   7   6   5   4   3   2   1

The paper used in this publication meets the minimum requirements of ANSI/NISO Z39.48–1992 (R 2002) (*Permanence of Paper*).

Cover image: *Suisun Marsh* (n.d.) by William Franklin Jackson. Crocker Art Museum Purchase.

CONTENTS

Color maps follow page 148

CONTRIBUTORS

JOSHUA T. ACKERMAN is a principal investigator with the U.S. Geological Survey, Dixon, California, and an associate in the Department of Wildlife, Fish, and Conservation Biology at the University of California, Davis. Contact: jackerman@usgs.gov

PETER R. BAYE is an independent consulting ecologist in California's Central Coast region. Contact: baye@earthlink.net

EDWARD BURNS is a land stewardship specialist with the U.S. Department of Agriculture, Natural Resources Conservation Service. Contact: Edward.Burns@md.usda.gov

MICHAEL L. CASAZZA is a principal investigator for the U.S. Geological Survey, Western Ecological Research Center. Contact: mike_casazza@usgs.gov

ESA K. CRUMB is an ecologist and GIS analyst at Wetlands and Water Resources, Inc. Contact: esa@swampthing.org

JOHN M. EADIE is the Dennis G. Raveling Professor of Waterfowl Biology and chair of the Department of Wildlife, Fish, and Conservation Biology at the University of California, Davis. Contact: jmeadie@ucdavis.edu

CHRISTOPHER ENRIGHT is senior water resources engineer, Delta Science Program, State of California. Contact: cenright@water.ca.gov

PEGGY L. FIEDLER is the director of the University of California Natural Reserve System in the UC Office of the President. Contact: peggy.fiedler@ucop.edu

BRENDA J. GREWELL is a research ecologist with the USDA-ARS Exotic and Invasive Weeds Research Unit and an associate in the Department of Plant Sciences at the University of California, Davis. Contact: bjgrewell@ucdavis.edu

MARK P. HERZOG is a wildlife biologist with the U.S. Geological Survey, Dixon, California. Contact: mherzog@usgs.gov

PATRICK R. HUBER is a conservation scientist at the University of California, Davis. Contact: prhuber@ucdavis.edu

MEGAN E. KEEVER is a botanist and plant ecologist with Stillwater Sciences. Contact: megan@stillwatersci.com

AMBER D. MANFREE is a PhD candidate in geography at the University of California, Davis. Contact: admanfree@ucdavis.edu

PETER B. MOYLE is a professor in the Department of Wildlife, Fish and Conservation Biology and associate director of the Center for Watershed Sciences at the University of California, Davis. Contact: pbmoyle@ucdavis.edu

TEEJAY A. O'REAR is a research scientist and coordinator of the long-term Suisun Marsh Fish Study at the University of California, Davis. Contact: taorear@ucdavis.edu

ARTHUR M. SHAPIRO is Distinguished Professor of Evolution and Ecology at the University of California, Davis. Contact: amshapiro@ucdavis.edu

STUART W. SIEGEL is president and principal wetland and environmental scientist at Wetlands and Water Resources, Inc. Contact: stuart@swampthing.org

ALISON N. WEBER-STOVER is a program associate at the Bay Institute. Contact: weberstover@bay.org

GREGORY S. YARRIS is the science coordinator for the Central Valley Joint Venture. Contact: GYarris@calwaterfowl.org

Suisun Marsh is one of the most remarkable natural areas in California. It is the largest tidal wetland in the state and perhaps on the Pacific coast. It not only serves as a refuge for a high diversity of native plants and animals, but is a major area of "open space" in an increasingly urban region. Yet Suisun Marsh may also be one of the least appreciated wetland ecosystems remaining in the state, despite its proximity to major urban areas. For the thousands of people who drive along it or pass through it every day, the Marsh is just a windy, vacant space, mysteriously without large buildings dominating the view. This attitude is changing, as reflected in the renewal of the waterfront in Suisun City on the Marsh edge, and in the increasing numbers of visitors to Rush Ranch, a property stewarded by the Solano County Land Trust as protected Open Space. The Marsh's value for nature-based recreation and as habitat for native plants and animals is only going to grow. Indeed, Suisun Marsh is looked upon more and more as a place for "restoration" and "mitigation" for habitats lost in other parts of the San Francisco Estuary, despite the Marsh's singular and distinctive nature.

Suisun Marsh has a long history of being managed for its wildlife, starting with the native peoples who harvested its natural bounty. Indeed, it exists today as a large wetland area mainly because of its value for waterfowl hunting. Over the past 120 years, hunters purchased much of the Marsh, successfully prevented industrialization of large wetland expanses, established wildlife refuges, and lobbied government agencies for various kinds of protections, including guarantees of freshwater inflow. Nevertheless, we think it unlikely that the Marsh will be managed in the future primarily for waterfowl, although providing habitat for ducks, geese, and hunters will always be part of its management. Instead, we envision that it will be managed more and more for native biodiversity, especially endemic species threatened with extinction, and for nature-based recreation. As shown in this book, the future Marsh is bound to be quite different from the present because of rising sea level, climate change, land-use changes, and changing human priorities, demands, and attitudes.

In this short book, we promote the idea of Suisun Marsh as a place of constant change throughout its entire history, but with the pace of change accelerating rapidly. We also promote the idea that the Marsh is not an island of habitat, but a node in a network of natural areas, both in the San Francisco Estuary and in surrounding hills and plains. This network is largely unrecognized but has huge potential for protecting the many endemic plants and animals that inhabit the region, provided that the connectivity among its pieces can be maintained or enhanced. All of this means that land and resource management decisions made (or not made) today in the region will have profound effects on the future functioning of the Marsh. The first eight chapters of this book demonstrate why this is so. The final chapter provides four possible future scenarios to illustrate landscape-scale results of different decision pathways.

The perception that large-scale change to Suisun Marsh is inevitable was the stimulus for a symposium on May 23, 2011, entitled "Suisun Marsh in the 21st Century: A Landscape of Change and Opportunity." The symposium, in turn, was a catalyst for this book, in which we look at the Marsh's past and present in order to examine its future in a 100-year time horizon.

The editors and authors of this book have long histories of studying and recreating on the Marsh. We are all deeply concerned about its future. While the Marsh is likely to remain a semi-wild place and a refuge for wildlife under almost any scenario, we would like to see choices made that maximize its ecosystem functions and take into account predicted shifts in ecological drivers. The Marsh should always be a center for native biodiversity in the San Francisco Estuary. Not surprisingly, this book is dedicated to that very premise, by summarizing what is known about its existing values to the native biota of the region and projecting what is likely to happen to this biota in the future under different management strategies. We try to be both idealistic and realistic in our approach. Our goal is to influence thinking about the future of Suisun Marsh so that it can be managed in a reasoned way, not just in response to crises. It is a wonderful, remarkable, dynamic place and we would like to keep it that way.

Peter B. Moyle, Amber D. Manfree, and Peggy L. Fiedler

ACKNOWLEDGMENTS

We thank our many colleagues who have roamed the Marsh with us and shared their knowledge of its flora, fauna, and ecology. The many enthusiastic and skeptical participants in the 2011 Suisun Marsh symposium collectively encouraged us to write this book. The symposium was sponsored by the Delta Science Program of the Delta Stewardship Council, the Center for Aquatic Biology and Aquaculture, and the Center for Watershed Sciences at the University of California, Davis. The Center for Watershed Sciences at UC Davis generously supported Amber Manfree so that she could devote her time to this book.

We also recognize the pioneering work of the late Randall L. Brown, who had key responsibility for the first Suisun Marsh workshop in March 2004 and whose summary of the workshop is still a useful source of insights and information (http://www.science.calwater.ca.gov/pdf/suisunmarsh_report_093004.pdf). His consistent promotion of independent science-based research in the Marsh is a model for other administrators to follow.

Special thanks to:

Alison Whipple, Chuck Striplen, and Robin Grossinger of the San Francisco Estuary Institute, as well as Joseph Honton and Bethany Hopkins, for advice and insight (chapter 2).

Jules Evens of Avocet Research Associates, Laureen Barthman-Thompson, Leonard Liu of Point Reyes Bird Observatory, Isa Woo of the U.S. Geological Survey, and Katherine Smith for information and review, and the California Department of Fish and Wildlife (Sara Estrella) for providing updated species faunal lists for Suisun Marsh (chapter 6).

Robert Schroeter, Bruce Herbold, Ted Sommer, Bill Bennett, Alpa Wintzer,

John Durand, Steve Chappell, and Orlando Rocha of the Suisun Resource Conservation District, and Kate Le and Terri Fong of the California Department of Water Resources, all of whom have contributed considerably to our understanding of Suisun Marsh's fishes and invertebrates (chapter 7).

The synthesis in chapter 5 was supported by the U.S. Geological Survey. We thank state and federal biologists, California Department of Fish and Wildlife (Dan Yparraguirre and Shaun Oldenburger), and U.S. Fish and Wildlife Service (Mike Wolder, Cheryl Strong, and Khristi Wilkins) for providing the data for the midwinter waterfowl survey, the parts survey, band recovery data, and the breeding population survey. Alex Hartman produced the maps of the western United States. Early versions of the manuscript were reviewed by Mike Miller, Phil Garone, Shaun Oldenburger, Steve Chappell, and Bruce Wickland. The use of trade, product, or firm names in the publication is for descriptive purposes only and does not imply endorsement by the U.S. Government.

1

Introduction

Peter B. Moyle, Amber D. Manfree, and Peggy L. Fiedler

Suisun Marsh has always been regarded as a remarkable place, made especially attractive by its abundance of fish, wildlife, and useful plants. The remarkable nature of Suisun Marsh stems from the coincidence of a number of factors.

Size. At about 470 km^2, the Marsh is often referred to as the largest brackish-water marsh on the western coast of North America (see box 1.1). While this claim is debatable, Suisun Marsh is certainly the largest such marsh in California.

Connectivity. Located in the middle of the San Francisco Estuary (see maps 1 and 2 in color insert), the Marsh is tightly linked to Suisun Bay. Both have strong interactions, through the movement of water and organisms, with both the Sacramento–San Joaquin Delta upstream and San Francisco Bay downstream. The Sacramento River delivers fresh water to the Marsh, and strong tides mix it with salt water pushed upstream from the Pacific Ocean, so the Marsh's salinity varies with place and season. From a terrestrial perspective, the Marsh is linked to both the Fairfield–Suisun City urban area and the watersheds above them, as well as to more natural areas to the north, including the Potrero Hills and the vernal-pool prairies (e.g., Jepson Prairie) that have further connections to the North Delta and the Yolo Bypass. On an even larger scale, the Marsh is an important stop on the Pacific Flyway, providing habitat for waterfowl and shore-birds migrating from the Arctic to the tropics, as well as for those wintering in central California.

BOX 1.1. DEFINING SUISUN MARSH

Brenda J. Grewell

Suisun Marsh is a well-marked entity on maps, yet its exact boundaries depend on who is defining it. Usually, the Marsh is defined by political boundaries, to exclude (among other things) urban areas and large areas of open water. The most commonly used definition is that found in the Suisun Marsh Habitat Management, Preservation, and Restoration Plan (or "Suisun Marsh Plan"; U.S. Bureau of Reclamation et al. 2011), which indicates total acreage as being roughly 100,000 acres, including 405 km² (52,000 acres) of managed wetlands, 104 km² (25,700 acres) of tidal bays and sloughs, 31 km² (7,700 acres) of tidal marshlands, and 68 km² (16,700 acres) of uplands, mostly lowland and hillside grasslands, although some agricultural and developed areas are included. The numbers are obviously rounded, which suggests that the actually boundaries are somewhat ill defined, despite being legal boundary lines. The Suisun Marsh Plan boundary, used by the Suisun Resource Conservation District for their planning, excludes some existing and historical tidelands because it accepts Highway 12 and other roads as boundaries rather than more natural watershed or elevational boundaries.

The San Francisco Bay Area Wetland Ecosystem Goals Project (1999) and the U.S. Fish and Wildlife Service Recovery Plan for Tidal Marsh Ecosystems, Northern and Central California, take a broader geographic approach. Both of these efforts set the eastern boundary of the Suisun region at approximately Broad Slough and extend the southern boundary to include all tidelands bordering Suisun and Honker bays. This means that Suisun Marsh includes Brown's Island and the extensive tidal marshlands on state, federal, and private lands on the Contra Costa shoreline, in contrast to the usual southern bound-

Biodiversity.   Suisun Marsh and the uplands around it, such as the Potrero Hills, support diverse habitats for both resident and migratory species. This habitat diversity is reflected in the large array of plants and animals, especially native species: 200+ plant species, 180+ bird species, 45 mammal species, 15 reptile and amphibian species, 60 species of butterflies, and 50+ species of fish. Some of these, such as Suisun song sparrow (*Melospiza melodia maxillaris*) and Suisun thistle (*Cirsium hydrophilum* var. *hydrophilum*), are largely endemic to the Marsh.

History.   Suisun Marsh has a long history of wildlife conservation, especially of waterfowl. Early duck-club owners successfully protected the Marsh from various schemes to use it for waste disposal or urban development. The Marsh exists today as a center of fish and wildlife diversity because of its continued management as a wetland ecosystem, albeit with a focus on waterfowl hunting. Most of

ary that follows the deep-water channel through the middle of Suisun Bay. Both reports also extend the regional western boundary of the Marsh into Carquinez Straits, therefore including the city of Benicia and Southampton Marsh/Bay. These areas are more similar to the central Suisun Bay region than they are to areas west of Dillon Point, such as the Napa River and San Pablo Bay. Also included, if vaguely, are the lower reaches of major streams, including Green Valley Creek, Sacramento River, Suisun Creek, and Walnut Creek. The total area encompassed by these documents is 360 km² (89,000 acres) of wetlands, channels, and bays and 91 km² (22,500 acres) of adjacent uplands, or 451 km² (111,500 total acres).

Just to confuse things further, according to the 1959 Delta Protection Plan, the legal boundary of the Delta includes a portion of Suisun Marsh. The western boundaries of this "legal" Delta include Montezuma Slough almost up to the salinity control gates, the Contra Costa shoreline of Suisun Bay to a line drawn through Honker Bay, Simmons Point–Chipps Island, and the shoreline well west of the city of Pittsburg.

In this book, we use the greater Suisun Marsh ecosystem as our reference rather than adopting the more limited circumscriptions used by Suisun Resource Conservation District or the Suisun Marsh Plan (see map 2 in color insert). But because most studies use the standard boundaries, we are often left with the statistics they generate. However, in discussing Suisun Marsh as an ecological entity, we are largely discussing a region that arbitrarily ends in the middle of Suisun Bay. This means that we have, for the most part, excluded discussion of the fringing marshes on the southern side of the Bay in northern Contra Costa County, partly because we know much less about these marshes than we do about the core Suisun Marsh. We consider, nevertheless, that many of the challenges they face in the future are similar to those of the main Marsh.

the land in the Marsh proper is either privately owned duck clubs or public space devoted to hunting and wildlife management.

Open Space.   Being on an urban fringe, the Marsh provides a major oasis of natural habitats in a rapidly developing area. The success of Rush Ranch Open Space as a nature center in the northeast quadrat of the Marsh demonstrates that it is increasingly important as open space for urban-weary humans. But its "vacant" nature also means ongoing competition to use it for other purposes, for example as a destination for toxic runoff or for disposal of wastewater from sewage treatment.

Research.   In part because Suisun Marsh is so accessible to several major universities, it is an important area for environmental research, as the chapters in this book illustrate. With the growing realization of the importance of estuarine wetlands (Batzer and Sharitz 2006; Silliman et al. 2009), especially along the San Francisco Estuary (Palaima 2012), Suisun Marsh has increasingly become a focal point for wetlands research, including studies related to ecological functions and wetland conservation under various sea-level-rise scenarios.

Not surprisingly, with all these attributes, Suisun Marsh is increasingly looked upon as a major conservation area and wetland mitigation site to compensate for future development throughout the San Francisco Estuary. Just how actions in Suisun Marsh can mitigate declines in natural habitat in other parts of the Estuary is not entirely clear. The Marsh is already intensively managed as wild lands, and large segments of it are likely to transition to "natural," brackish-water, intertidal wetlands, or to open water, given the inevitability of sea-level rise. Whatever the reasons, it is clear that the Marsh will have increasing value to society as a protected natural area as time goes on, despite its history of constant change. In the past 200 years alone, it has been transformed from a natural tidal marsh used as a bountiful source of food and fiber by local Miwok people to farmland and pasture, and then to highly managed waterfowl habitat with over 350 km of dikes that separate marshlands from intervening sloughs. Management of most of these duck clubs and wildlife areas today focuses on producing resident mallard and attracting migratory waterfowl in winter. Yet increasing numbers of protected areas are maintained as habitat for rare species such as salt marsh harvest mouse (*Reithrodontomys raviventris*), reflecting future changes in land management.

At the same time, the Marsh is continually invaded by alien species, heralded by the perennial pepperweed (*Lepidium latifolium*) that often dominates dikes and other upland areas, and by the giant reed (*Arundo donax*) that is moving into marshy areas. Aggravating these invasions, duck clubs are often managed for alien plant species to provide food for ducks. New species of fish and aquatic invertebrates invade on a regular basis, such as shimofuri goby (*Tridentiger bifasciatus*) and Siberian prawn (*Exopalaemon modestus*). Meanwhile, native fishes such as delta smelt (*Hypomesus transpacificus*) decline. Novel assemblages of organisms, featuring species from all over the world, have been formed and are constantly changing as new species arrive and native species decline (some to extinction), presenting new challenges for managers. As discussed in chapter 9, Suisun Marsh is a dynamic ecosystem, requiring new, adaptive approaches with firm, widely agreed-upon goals for what our society wants the Marsh to be like in the future.

Change will continue in the Marsh at an accelerated pace. Most conspicuously, sea level is rising. Best estimates are that it will rise by 0.8–1.5 m by 2100, with high tides, storm surges, and floods pushing the water higher (Cayan et al. 2009; Knowles 2010). It is widely accepted that most present estimates of sea-level rise are conservative (Rahmstorf et al. 2012) and that occasional megafloods can be expected from extreme precipitation events (Dettinger and Ingram 2013), so change may be unexpectedly fast. Meanwhile, sediment supplies from inflowing rivers are diminishing, while subsidence in nontidal areas due to specific land-management practices is continuing apace. Given that much of the Marsh is at or below sea level, large portions are likely to become permanently open-water subtidal habitat, as dikes give way or are breached in the not-so-distant future.

Planning efforts for the future Marsh are under way, but the Suisun Marsh Habitat Management, Preservation, and Restoration Plan (or "Suisun Marsh Plan"; U.S. Bureau of Reclamation et al. 2011) focuses on maintaining the Marsh in its present state as much as possible, even as lands subside and increasingly higher tides spill over existing dikes. This status quo focus may be possible within a 30-year time frame, especially for areas of higher elevation. But as time goes on, the contemporary ecological conditions and processes will become less and less sustainable, in particular because one key emphasis of the Suisun Marsh Plan is to dredge tidal channels, a management practice certain to exacerbate the effects of sea-level rise (chapters 3 and 8).

So what is the long-term future of Suisun Marsh? As the analyses presented in this book illustrate, there is no doubt that Suisun Marsh will continue to function as open space in an increasingly urbanized environment and as important habitat for wildlife, fish, and plants (although salt-tolerant species will become increasingly important) for at least the next century. The distribution and severity of such changes will depend on the rate of sea-level rise and our response to it. Under one long-term scenario, tidal water will eventually flood the streets of Suisun City (Knowles 2010). However, a Marsh that contains more tidal habitat than at present is more likely to help buffer urban areas from tidal flooding.

To some degree, the direction and extent of future change to Marsh habitats can be managed, as we explore in four scenarios in chapter 9. One future, for example, is to continue to manage the Marsh as a series of independent, "postage stamp" units, as is largely done today with the 158 private duck hunting clubs and the handful of public wildlife areas. In this future, the best waterfowl hunting habitat likely will belong to whoever builds the highest and strongest dikes or begins with the highest-elevation land along the Marsh fringes.

Another potential scenario generally envisions the Marsh as one area with unified wetland management. For example, one strategy might be to construct the biggest dike possible along some strategic line of defense across the Marsh, protecting marshlands behind it and, eventually, urban areas from saltwater incursion. This, of course, is what has been done in areas like the Netherlands and in New Orleans, at enormous expense. From an ecological point of view, a more desirable scenario (and one more likely to succeed) is to work *with* sea-level rise and other change stressors to create a "new" marsh with as many desirable features as possible (e.g., natural tidal channels). This latter scenario, of course, would require regulatory agencies, local landowners, and other California citizens to agree upon just what features are the most desirable.

FIGURE 1.1. Two oblique aerial views of Suisun Marsh from ca. 1909 (left) and 2006 (right), looking south along Suisun Slough. Note that while major slough structure appears to be similar, many sloughs have been channelized or cut off by dikes from the main sloughs. (1909 photo courtesy of Solano Historical Society. 2006 photo by Peter B. Moyle.)

The scenario that comes the closest to fitting reality in 100 years or so will depend on how well existing information and projections of environmental change are used. This book is designed to provide an introduction to these subjects to start the process of rethinking the future of Suisun Marsh, much as Lund et al. (2007, 2010) have done for the Delta. In the following chapters, we

- review the ecological history of Suisun Marsh;
- describe, to the best of our knowledge, the present-day Marsh and how it functions;
- describe the status and trends of the biota, focusing on plants, fish, waterfowl, and terrestrial vertebrates;
- describe the forces of change for the Marsh, especially sea-level rise; and
- examine alternative futures for the Marsh.

We realize that our approach omits much information present in the previous plans, reports, and publications about the Marsh. However, we anticipate that this book will be most useful if it presents a broad overview reflecting new information on the Marsh and new syntheses of this information. Our goal is to encourage managers to look for and implement a plan for a "soft landing" for the Marsh ecosystem, rather than waiting for a hard crash caused by rapid environmental change. Our general attitude is that of reconciliation ecology (Rosenzweig 2003), which treats humans as an integral part of every ecosystem and recognizes that conservation will work best if this truism is taken into account. We recognize that the Marsh has been altered by humans since the beginning of its 6,000-year history (chapter 2), yet it retains many of its fundamental characteristics as a tidal system of sloughs and marshlands (figure 1.1). Regardless of how the Marsh changes in the future, it will continue to be a major area of open space, tidal sloughs, and marshlands, as well as important habitat for fish and wildlife in an increasingly urbanized landscape.

REFERENCES

Batzer, D. P., and R. R. Sharitz, eds. 2006. Ecology of Freshwater and Estuarine Marshes. Berkeley: University of California Press.

Cayan, D., M. Tyree, M. Dettinger, H. Hidalgo, T. Das, E. Maurer, P. Bromirski, N. Graham, and R. Flick. 2009. Climate change scenarios and sea level rise estimates for the California 2008 Climate Change Scenarios Assessment. Draft paper, California Energy Commission, CEC-500-2009-014-D.

Dettinger, M. D., and D. L. Ingram 2013. The coming megafloods. Scientific American 308(1): 64–71.

Knowles, N. 2010. Potential inundation due to rising sea levels in the San Francisco Bay region. San Francisco Estuary and Watershed Science 8: 1–19. Lund, J., E. Hanak, W. Fleenor, W. Bennett, R. Howitt, J. Mount, and P. B. Moyle. 2010. Comparing Futures for the Sacramento–San Joaquin Delta. Berkeley: University of California Press.

Lund, J., E. Hanak., W. Fleenor, W., R. Howitt, J. Mount, and P. Moyle. 2007. Envisioning Futures for the Sacramento–San Joaquin Delta. San Francisco: Public Policy Institute of California.

Palaima, A., ed., 2012. Ecology, Conservation, and Restoration of Tidal Marshes: The San Francisco Estuary. Berkeley: University of California Press.

Rahmstorf, S., G. Foster, and A. Cazenave. 2012. Comparing climate projections to observations up to 2011. Environmental Research Letters 7: 044035.

Rosenzweig, M. L. 2003. Win-Win Ecology: How the Earth's Species Can Survive in the Midst of Human Enterprise. Oxford: Oxford University Press.

Silliman, B. R., E. D. Grosholz, and M. Bertness, eds. 2009. Human Impacts on Salt Marshes: A Global Perspective. Berkeley: University of California Press.

U.S. Bureau of Reclamation, U.S. Fish and Wildlife Service, and California Department of Fish and Game. 2011. Suisun Marsh Habitat Management, Preservation, and Restoration Plan. Final environmental impact statement/environmental impact report. U.S. Bureau of Reclamation, Sacramento.

GEOSPATIAL DATA SOURCES

CalAtlas. 2012. California Geospatial Clearinghouse. State of California. Available: http://atlas.ca.gov/. Accessed: 2012.

Contra Costa County. 2013. Contra Costa County Mapping Information Center. Available: http://www.ccmap.us/. Accessed: January 2013.

Gesch, D., M. Oimoen, S. Greenlee, C. Nelson, M. Steuck, and D. Tyler. 2002. The National Elevation Dataset. Photogrammetric Engineering and Remote Sensing 68: 5–11.

Solano County. 2013. Geographic Information Systems Homepage. Solano County Department of Information Technology. Available: http://www.co.solano.ca.us/depts/doit/gis/. Accessed: November 2012.

U.S. Geological Survey. 2013. National Hydrography Dataset (Watershed Boundary Dataset). Available: http://nhd.usgs.gov/wbd.html. Accessed: November 2010.

2

Historical Ecology

Amber D. Manfree

North of Suisun bay, stretching in land for many miles, is a low flat country threaded by tortuously winding sloughs in which the tide waters fall and rise and overflow into numerous ponds.

—WILLIS LINN JEPSON, 1905

Suisun Marsh is a place of constant and relatively rapid change, with vital connections to regional ecological processes. It is not a place of stasis or isolation. One of the first things to become apparent when looking at historical maps is how natural forces such as faulting, winds, tides, and river flows have sculpted the landscape. Telltale signs of geomorphic processes are now largely obscured by human modifications; nevertheless, by comparing maps made at different times, one can see drivers of change in the patterns that emerge. Novelties and subtleties of landscape ecology are often found in historical accounts, and tracking land use through time can show both likely current and future management outcomes. In this chapter, we first review forms and functions in the Marsh landscape and then examine landscape changes wrought by humans.

GEOLOGY, TECTONICS, AND GEOMORPHOLOGY

About 25 mya[1], the Pacific Plate met the North American Plate and began to slide north-northwestward. Ten million years later, this transformational boundary reached the area that would become the San Francisco Estuary. By 3.5 mya, contact between plates changed from plates sliding past one another to plates sliding at an angle to one another, a combination of sliding and compression

1. Mya = millions of years ago.

termed *transpression*. The resulting increase in friction initiated the rise of the Sierra Nevada. About 2 million years later, these same forces began to lift the Coast Ranges that frame the San Francisco Estuary we know today. The Coast Ranges on the North American Plate were pulled and stretched by the Pacific Plate in a north-northwesterly direction in relation to the Central Valley, a process that continues today. Although all of the Coast Ranges move in the same general direction, westward segments move relatively faster than eastward ones. Thus, the San Francisco Estuary landscape has been, and continues to be, rapidly reconfigured—at least in geologic time (Sloan 2006).

The Estuary is geologically young, having changed dramatically over the past 1 million years. The Central Valley of 1 mya was a large inland lake, and what is now the floor of San Francisco Bay was a plain crossed by small rivers. Around 620,000 years ago, this configuration shifted. The inland lake began draining to the Pacific by down-cutting the present Sacramento River channel that passes by Suisun Marsh and then flowing through Carquinez Strait and into San Francisco Bay. Much later, during the ice age 20,000 years ago, sea level was lower and the ocean shoreline was situated nearly 50 km (30 mi) west of the Golden Gate. At this time, California's greatest river meandered between floodplains that now lie beneath San Pablo and San Francisco bays.

The valley in which Suisun Marsh formed was created by tectonic activity along the Concord–Green Valley fault system, as well as by smaller nearby faults. A series of northwest–southeast trending anticlines and synclines undulate to depths up to 2,500 m beneath the peat soils and alluvium of the eastern Marsh and Grizzly, Suisun, and Honker bays. These anticlines and synclines are underlain by southwest-dipping thrust faults (Unruh and Hector 2007). The Potrero Hills are Tertiary rocks uplifted by an anticline; the Montezuma Hills are made of younger alluvial deposits that also have been uplifted (Sloan 2006).

Approximately 10,000 years ago, sea level was still much lower than it is today. At this time, the Suisun Marsh region was a river valley. Sea level rose as glaciers melted in the earth's warming climate and fringing marshes advanced upslope as the bays filled with water (Atwater et al. 1979). At times, the water rose so fast that areas of marsh did not keep pace and were drowned, as seen in layering of marsh deposits and intertidal and subtidal sediments (Atwater et al. 1977). Around 6,000 years ago, the rate of sea-level rise slowed dramatically, fostering the formation of extensive tule (*Schoenoplectus* spp. and *Bolboschoenus* spp.) marshes in the eastern Estuary, including Suisun Marsh (Atwater et al. 1979; Malamud-Roam and Ingram 2004; Drexler et al. 2009). As rising waters filled the lower drainages of Suisun Valley Creek and its tributaries, tules grew and decayed and sediment was deposited by rivers, tides, and wind, forming a thick layer of peat soils.

CLIMATE VARIATION IN RECENT MILLENNIA

Earth's climate has warmed slowly since the last ice age (Burroughs 2007). In the San Francisco Bay region, the change has not been a steady, gradual shift. Sediment cores and tree rings reveal wide fluctuations in climatic conditions occurring on time scales of a few years to decades. This erratic variation can be attributed to the climatic boundary zone in which the region is situated, with wetter, cooler conditions to the north and drier, hotter conditions to the south. In any given year, weather may be moderate or more like that to either the north or south. Paleoclimate studies have revealed relatively recent extended droughts and deluges unlike any seen since Euro-American colonization of California. Compared to the past 2,000 years, the past 150 years have exhibited an anomalous pattern of very stable conditions (Malamud-Roam et al. 2007). The historical record demonstrates that unpredictable shifts and conditions less favorable than those to which we are accustomed are typical. Imminent effects of anthropogenic climate change further increase our uncertainty in the future climate of the San Francisco Bay region.

PROCESS-DRIVEN SUBREGIONS

Discussions of Suisun Marsh often refer to the "western Marsh" and "eastern Marsh," and numerous reports have carved the Marsh into alternative subregions to suit different purposes. For example, the Suisun Marsh Habitat Management, Preservation, and Restoration Plan (U.S. Bureau of Reclamation et al. 2011) divides the Marsh into four subregions for management purposes (see chapter 9). Here, subregions are delineated according to physical drivers of landscape formation to understand processes contributing to the array, extent, and proportion of habitat types within the Marsh.

Modifications that transformed geomorphic processes within Suisun Marsh did not begin until the late 1800s. Levees and dikes, hydrologic diversions, and other such infrastructure came later to Suisun Marsh than to the Sacramento–San Joaquin Delta or the urban edges of San Francisco Bay. Surveying and mapping marshes was a difficult task in the era before aerial photography; consequently, our map record contains considerable ambiguity. Yet maps made between 1875 and 1910 tell a compelling story about physical dynamics prior to management overhauls made by settlers. Sufficient geomorphic features are represented to establish functional zones that likely existed and that still affect the landscape.

A careful look reveals at least four major and three peripheral process-driven subregions in the historical Marsh, with major subregions being those that cover large contiguous areas. Major subregions include the (1) western Marsh, including Suisun and Goodyear sloughs and their tributaries; (2) southeastern Montezuma Slough distributary network; (3) northeastern Nurse–Denverton slough area; and (4) Southern marsh. Peripheral subregions are associated with marsh edges and include transitional landscape features covering smaller areas such as the (5) marsh–upland transition zone, (6) ephemeral wetlands near the marsh edge, and (7) tidal mudflats (see map 3 in color insert). In addition to the subregions outlined above, there are finer-scale subregions (e.g., marsh plain microtopography) as well as complex physical processes (e.g., slough hydrodynamics) with important ecological implications that are beyond the scope of this chapter. Upland ecosystems, though intrinsically linked to marsh processes, are likewise not evaluated here.

The Western Marsh: An Estuarine Gradient

On the western side of Suisun Marsh, Suisun Slough and its tributaries follow the classic marsh pattern observed where freshwater creeks meet tidal bays. Not including low-lying marsh, the Suisun Slough watershed covers approximately 486 km² (120,000 acres). The processes and ecology of this Marsh subregion might be compared to nearby creeks with similar watershed area and topographic relief, such as Sonoma Creek in Sonoma County. Suisun, Goodyear, Cordelia, and smaller sloughs drain Green Valley, the eastern flank of the Sulphur Springs Hills, and the western side of the Potrero Hills. Western Marsh sloughs are highly sinuous, which suggests they have been developing for a long time (Atwater et al. 1979; Hall 2004).

The boundary of the historical western Marsh crosses Joice Island from Grizzly Bay to the Potrero Hills. The boundary is imprecise because of continuous tidal and fluvial fluctuations that affect the direction of flow in Cutoff Slough, which connects Suisun Slough to Montezuma Slough at the northern end of Joice Island. The historical Suisun Slough followed an estuarine gradient, with fresher conditions upstream, where it is fed by Suisun and Green Valley creeks, to increasingly brackish conditions downstream, except when freshened at the downstream end by very high Delta outflows. The lower western Marsh was the most tidally influenced and least river-influenced subregion within Suisun Marsh. Reconstructions of historical salinity provide evidence that Suisun Marsh's western subregion was generally the most saline part of the Marsh (Collins and Grossinger 2004), a condition driven by proximity to San Pablo Bay and distance from the Delta. Conditions varied widely from year to year, however, depending on Delta outflows.

Maps show that the historical western Marsh featured numerous ponded water bodies scattered across the marsh plain among channels, a feature common to tidal marshes around San Francisco and San Pablo bays. These ponds primarily appear west of Suisun Slough between Wells and Cordelia sloughs.

U.S. Geological Survey (USGS) topographic maps delineate 87 ponds in the marsh plain of western Suisun Marsh around the turn of the 20th century (USGS 1896, 1918a, 1918b, 1918c). Other early maps indicate additional natural ponds near Cordelia Slough (Stoner 1934, 1937) and reaffirm that there were many ponds in this area (Arnold 1996), although pond numbers and locations represented in maps vary.

Ponds in the western Marsh are an integral part of local duck-hunting lore, and these natural features inspired environmental management decisions made throughout the 20th century. In 1900, only a few duck hunting clubs could be found in the Marsh, and hydrologic management was nascent, so ponded water features on maps drawn at this time were likely natural. They were clearly habitat for migrating flocks of waterfowl, as attested by market hunters' stories of the mid-1800s, the locations of the earliest hunting clubs, and the route of the railroad that serviced them (Arnold 1996).

Numerous geomorphic processes can lead to the formation of ponded areas in marshes. *Marsh pannes* form where water is hydrologically trapped and isolated on the marsh plain and is so deep and persistent, or so saline because of evaporation, that plants cannot colonize. *Ponds* on the marsh plain are connected, at least occasionally, to tidal flows and are therefore fresher than pannes. *Sag ponds* are formed by subsidence along faults and can occur in uplands or in marshes, and *salinas* are shallow, hydrologically isolated, periodically dry wetlands that form in the marsh–upland transition zone and precipitate salts (Collins and Grossinger 2004).

The Cordelia Fault is located in the vicinity of the western Marsh and may have been a factor contributing to pond formation. The Sulphur Springs Hills shelter the western Marsh from prevailing winds to some extent, and this protection may have aided pond formation and affected duck habitat preferences (Monda and Ratti 1988), especially in comparison to the windblown southeastern and northeastern subregions of Suisun Marsh (see below).

Animals forage, trample, and puddle in wetlands, grazing on vegetation and excavating channels and ponds in the process (Mitsch and Gosselink 2007). Biogenic pond formation may have been driven by elk and other ungulates until their rapid decline in the 1850s, but this would not explain the skewed spatial distribution of ponds. Several early accounts credit the ducks themselves for pond formation in western Suisun Marsh. An early article by the renowned 20th-century California botanist Willis Jepson mentions duck clubs that were presumably located in the western Marsh, because that was the only area with clubs at the time. Jepson (1905) described interactions between ducks and landscape processes as "[w]hen the [wild] geese are in great numbers they eat out the tule so effectively that ponds, often of considerable extent, are formed." He explained that the widespread and prolific fennel (sago) pondweed (*Stuckenia pectinata*) subsequently colonized ponded areas, which were 3 to 4 ft deep. "Before the ducks come the ponds are filled with regular masses of the fennel pondweed, as even in its growth as a field of young grain." He is likely referring to ponds at the Ibis Gun Club, because it had ponds of this depth (Stoner 1937).

James Moffit (1938) also described sequential, species-specific waterfowl feeding activities that formed and then deepened the ponds:

> Originally the Lesser Snow Geese (*Chen hyperborea hyperborea*) made the ponds on this marsh by tearing up clumps of three-square (*Scirpus americanus*) to secure its bulbs for food . . . Then, the Whistling Swans (*Cygnus columbianus*), working in the areas opened by the geese, deepened the ponds to three feet or more by tilting up like surface-feeding ducks and reaching down with their long necks. Plant growth, of which sago pondweed (*Potamogeton pectinatus*) is by far the most important one locally, becomes established when ponds with proper conditions of salinity and requisite depth (18 inches or more) are created. Sago pondweed, an excellent food plant, attracts surface-feeding ducks, notably Pintail (*Dafila acuta*), until the ponds are deepened so that the growth is no longer within reach of the surface-feeders. The ponds then become attractive to diving ducks, of which the Canvasback (*Nyroca valisineria*) is the only common one in this region. Canvasbacks in their feeding operations, may further deepen the ponds.

Annual precipitation totals during the late 1800s and early 1900s were frequently above average, so large Delta outflows kept all of Suisun Marsh unusually fresh during this period. Saltier conditions may have contributed to pond formation in previous periods of extended drought. However, given that sago pondweed thrives in brackish and fresh water and was reliably present, well distributed, and prolific in ponds in mid- to late summer when ducks stopped through on their migrations, pond water was not hypersaline at the driest time of year in the western Marsh and, thus, persistent ponding was not due to hypersaline conditions. The historical role of waterfowl in pond creation and maintenance is still a bit of a mystery. Speculations on biogenic pond formation hinge on historical waterfowl abundance and use of ponds by waterfowl and other Marsh denizens.

The Southeastern Marsh: A Distributary Network

Suisun Marsh is sometimes referred to as "the western Delta," and its southeastern portion, bounded by the Montezuma Hills to the east and Montezuma Slough and adjacent marshes to the north, is the part that most lives up to that name. It is more similar to the Delta than to smaller estuarine systems around San Francisco Bay, with large freshwater inputs, webs of interconnected sloughs, and numerous islands. Historically, few ponds were located here. Wind-influenced accretion and disturbance were, and still are, major factors shaping this subregion.

The watershed that feeds the upstream end of Montezuma Slough at Van Sickle Island is that of the entire Sacramento–San Joaquin River system, which

BOX 2.1. HOW MUCH FRESHWATER INFLOW
IS NORMAL FOR SUISUN MARSH?

In the 1960s and 1970s, many water policy decisions were made regarding major dams and diversions of water from the Sacramento and San Joaquin watersheds. For the most part, these rules still apply. At the time the rules were set, the climate record in California did not extend much farther into the past than the years in which the first duck clubs were founded. Thus, the drought of the 1930s was considered an aberration in the context of the relatively wet periods observed before and after. The lack of understanding of California's long-term climate variability has had major impacts on water management practices in Suisun Marsh, because unusually fresh conditions were considered normal when water quality standards were set. In fact, the historical Suisun Marsh repeatedly experienced long periods of low freshwater inflow, presumably responding by becoming more like a salt marsh in areas closest to the bays. This variation would have greatly influenced how the landscape was used by waterfowl, fish, and other residents. According to the climate record of the past few thousand years (see chapter 4), Suisun Marsh almost never functioned as a freshwater marsh. Upstream water diversions have affected salinity, but arguably the change is similar to that caused by droughts in the past.

drains about 162,000 km^2 (40 million acres), or 40% of the state of California. After flows are pinched between the Montezuma Hills and the northern face of the Mount Diablo range, deltaic processes resume in the southeastern subregion of the Marsh as great rivers and tides intermingle. This geomorphic extension of the Delta is characterized by numerous islands (e.g., Chipps, Wheeler, Simmons, Grizzly, and Ryer) dissected by open-ended sloughs. Island-building processes are ongoing within the Sacramento–San Joaquin river channel. Shoals form where water moves slowly and channels are regularly dredged to maintain passage for ships, necessary even though sediment inputs are currently limited by upstream dams. The southeastern Marsh was historically fresher, on average, than the western Marsh (San Francisco Estuary Institute [SFEI] 2012).

While sediment delivered by the Sacramento and San Joaquin rivers played an important role in forming the entirety of Suisun Marsh, marsh-building sediment delivery has been greatest in the southeastern subregion in the recent past. The influx of sediment from hydraulic gold-mining operations in the late 1800s and early 1900s (see chapter 3) greatly accelerated sediment delivery and deposition. A large quantity of this sediment settled out of the slow-moving, wind-affected shallow water at the northeastern end of Grizzly Bay. The bayward portion of Grizzly Island accreted so rapidly that vascular plants were not a major part of the soil-building process; thus, the soils are mineral, and this area is not at risk of subsidence through oxidation of peat as are other areas of the Marsh (see box 2.3). Mercury used in gold-mining operations is likely present in these soils, a subject ripe for further study.

The near absence of ponds represented in the southeastern and northeastern Marsh subregions by early USGS topographic maps is striking, with only seven illustrated in these subregions combined. Comparing the first USGS (1896) topographic surveys of the western Marsh to the first surveys of the southeastern Marsh (USGS 1918a, 1918c), the latter maps are of overall better quality and almost no ponds appear in them. Absence of ponds in the southeastern Marsh is affirmed in early surveyor maps (Solano County Surveyor, 1920s).

Hydrology dominates landscape processes in the southeastern subregion of Suisun Marsh, but the wind also makes its mark. The atmosphere above the Montezuma Hills is among the most turbulent in the state, with mean annual wind speeds of up to 8.5 m/s (19 mph) (AWS Truepower 2010). Air masses are generally funneled through the Golden Gate and Carquinez Strait, rushing from the southwest toward the Central Valley, pushing everything on the landscape surface toward the northeast. Embayments on Frost Slough, on Little Honker Bay in the northeastern Marsh, and at the confluences of Nurse and Denverton sloughs (also in the northeastern Marsh) all appear at the downwind terminus of stretches of water that align with prevailing winds. Grizzly and Honker bays also follow this directional pattern. Further, it is likely that this wind disturbance prevents plants from colonizing and stabilizing shallow mudflats that would otherwise be prime territory for marsh building. Winds cause water to pile up in these shallow bays, an effect referred to as "seiches" or "wind tides." Material in the water collects along upwind banks, where currents are slower and water churns rather than washing through as in channels. Fine particles remain suspended, but heavier particles settle. In the case of Grizzly Bay, wind effects were probably a factor in rapid sediment accumulation following the Gold Rush. Prior to active flood management, seiches would have contributed to flooding of the marsh plain in high-water events much as they did in the Delta (Thompson 1957).

Wind has important effects on salinity gradients, sediment transport, food availability in the water column, and vegetation. Salt water is denser than fresh water and tends to run along the bed of a channel underneath outflowing fresh water. When wind mixes the water, there is less stratification between salty and fresh water, which prevents salty water from moving as far upstream as it would in the absence of wind. If the air is calm while salt water travels upstream, higher salinities will persist even after the wind picks up again (Lacy and Monismith 2000). Wind disturbance increases turbidity, which shortens the distance light penetrates the water column. The turbulence generated by wind also circulates nutrients and plankton in the water column. Excessive wind is a stressor on vegetation and can have effects ranging from hardening of exterior plant surfaces and stunted growth to no growth at all where winds are too intense. The winds passing over Suisun Marsh doubtless have major impacts on ecological processes on a number of levels.

The Northeastern Marsh:
Hydrologic Isolation Heightens Productivity

The Nurse–Denverton complex in the northeastern part of Suisun Marsh occupies only about 10% of the Marsh (24 km^2 or 5,900 acres), yet it is one of the most interesting locations because of its unusual geomorphic features and relatively high aquatic productivity. Although the watershed for this subregion is comparatively very small, with only about 65 km^2 (16,000 acres) of low-relief upland, in contrast with 486 km^2 (120,000 acres) of upland with moderate relief contributing runoff to the western Marsh, channels are well developed and relatively broad and deep. Partly because of its small watershed, hydrologic circulation is limited. Wind effects and hydrodynamics in Montezuma Slough also boost hydrologic residence time in this subregion, creating conditions conducive to a rich aquatic food web.

The larger sloughs in the Nurse–Denverton area cannot have been carved by the diminutive tributaries present today. Contemporary watersheds are small in size and lack topographic relief, so they could not deliver large or fast-moving inflows of water that would create broad, deep sloughs. Similarly, present-day tidal action is not powerful enough to carve large sloughs here. It is likely that the ever-shifting Coast Ranges were once configured in such a way that Nurse Slough was an outlet for, or deltaic branch of, a large river. The lower Napa River similarly may have been shaped by an ancient waterway on the scale of the contemporary Sacramento River (Loeb 2011). The drainage divide between the Nurse–Denverton area and the Delta is only approximately 3 m above sea level and is bordered on the north and south by hills uplifted in recent geologic history.

Landscape features suggest that wind also is a major geomorphic driver in the Nurse–Denverton subregion. The gap in the Coast Ranges between the Vaca Mountains and Mount Diablo functions like a trough, channeling air over Suisun Marsh and the low-lying Montezuma Hills to the Lindsey–Cache slough area in the Delta. Wind disturbance is probably what prevents marsh plants from colonizing shallow mudflats in oddly shaped Little Honker Bay, located at the end of the long, straight, southwest–northeast trending portion of Nurse Slough. Wind may also have been a factor in prevention of natural pond formation here because wind disturbance may have deterred ducks from visiting (Monda and Ratti 1988) and puddling for extended periods.

Regardless of the process by which the sloughs were formed, their current configuration has major ecological implications. Small watershed size limits the amount of storm discharge that might flush through the slough network. Nurse

Slough meets Montezuma Slough at nearly a right angle (see map 2 in color insert), and when incoming tides are moving rapidly or when outgoing winter flood waters are high, flow rushes through Montezuma Slough so powerfully that the water itself acts like a dam. Because of its momentum, water tends to push directly forward (northwest) rather than turning sharply to the northeast and flowing into the Nurse–Denverton region. This fast-moving water simultaneously obstructs slow-moving water at the mouth of Nurse Slough from flowing out into Montezuma Slough. Seiches also affect physical processes here, pushing water toward the northwestern ends of long fetches away from the system's outlet at the mouth of Nurse Slough.[2]

Hydrologic isolation and related high hydrologic residence time resulting from the physical processes described above have persisted into the present, although effects of upstream water management, especially the salinity control gates, keep the area fresher and dampen natural fluctuations in temperature and other water-quality parameters. Today, water quality in the Nurse–Denverton subregion is the least variable and is among the most consistently fresh in Suisun Marsh (O'Rear and Moyle 2009). This subregion has some functional similarities to standing bodies of water where lack of through-flow promotes stable conditions and retention of nutrients. Cattle grazing in adjacent uplands may be increasing nitrogen and sediment inputs throughout the Marsh, effects that would have proportionally greater impacts in the Nurse–Denverton complex than in other areas of the Marsh because of the lack of flushing. Fresher conditions and less fluctuation in water quality create a less stressful environment overall, one favorable to aquatic species that are less able to adjust to broad swings in environmental conditions.

Wind-driven hydrologic mixing increases dissolved oxygen and can boost productivity by distributing nutrients and food throughout the zone of mixing. Wind-driven mixing also keeps fine sediment suspended in the water column, contributing to the muddy appearance of Suisun Marsh's waterways. This turbidity is an abiotic aspect of habitat from which some aquatic species benefit because a limited line of sight provides cover from visual predators, such as striped bass and birds, and security for predators that rely on senses other than vision, such as sturgeon. Suspended sediment also mutes light penetration, which slows phytoplankton growth.

Compared to other sloughs in Suisun Marsh, Nurse and Denverton sloughs have been observed to have relatively high diversity and abundance of fishes

2. Effects of seiches during flood events would have been strong prior to levee construction in the Nurse–Denverton complex, when floods caused water levels to rise and cover the marsh plain, creating excellent fish nursery conditions.

BOX 2.2. RISKS OF IMPROVING CONNECTION

People often perceive increased hydrologic connectivity to be beneficial, for one reason or another. There have thus been several occasions when plans were made to connect Denverton Slough to Lindsey Slough in the Delta, which is only about 11 km (6.5 mi) distant by land. Some plans proposed extending Calhoun Cut, a canal constructed as part of an elaborate land speculation scheme in the early 1900s, to the upstream end of Denverton Slough. The expense of construction and concern that managing the location of the freshwater–saltwater boundary in the Delta would become more difficult have prevented such a project from being undertaken. In addition to effects on the freshwater–saltwater boundary, numerous ecological changes would cascade from the connection of Lindsey Slough to Denverton Slough. These include (1) creation of a new pathway for aquatic organisms to move between areas that are currently distant in water-miles; (2) increased hydrologic circulation and, therefore, lower hydrologic residence time in the Nurse–Denverton complex and at the current upstream end of Lindsey Slough; (3) an additional source of freshwater input to Suisun Marsh when water in the Sacramento River system is high; (4) less resistance against intrusion from tide pulses in the Nurse–Denverton area; and (5) changes in salinity that would shift plant community composition and diversity levels (Balling and Resh 1983).

(O'Rear and Moyle 2009). Denverton Slough stands out as an example of high-functioning brackish aquatic habitat. It is a moderately sized waterway with concomitant channel complexity that has not been diked off at the mouth from larger downstream sloughs (as seen in Volanti Slough), although it is worth noting that upper Denverton Slough was straightened in the early 1900s to improve passage for boats carrying agricultural freight. For this reason, as well as the presence of shoreline dikes, managed wetlands on the marsh plain, and grazing on adjacent uplands, the area is highly altered by any measure. Nonetheless, the Nurse and Denverton slough aquatic and wetland ecosystems support a high diversity and abundance of fish and other organisms. A more thorough evaluation of ecological conditions here could inform wetland conservation projects in the region.

The potential of the northeastern Nurse–Denverton subregion to accommodate sea-level rise should not be overlooked. Sloughs and marsh plains in this area are bordered by sparsely developed, low-rolling hills. Parcels are relatively large, and well-placed protected areas could provide areas for marshlands to move up-gradient in response to rising waters. At the same time, terrestrial habitat corridors from Suisun Marsh to Jepson Prairie and the northern Delta could be established.

BOX 2.3. THE SOUTHERN MARSH:
FRINGING CALIFORNIA'S GREAT RIVER

The fringing marsh on the Contra Costa County shore of Suisun Bay is some-times considered part of Suisun Marsh, although it is not treated as such in this book. It is worth reviewing briefly because it provides a glimpse of what might have been an alternative future of the Marsh had it not been adopted by water-fowl hunters.

The Contra Costa fringing marsh occupies a ribbon of shoreline between Martinez and Pittsburg, where the southern bank of the Sacramento–San Joaquin Delta meets the northern foot of Mount Diablo. The watershed for this subregion includes Walnut, Mount Diablo, Upper Alameda, and Marsh creeks, along with numerous smaller creeks and sloughs. At nearly 650 km² (160,000 acres), the watershed of this strip of marsh is comparable in size to that of Suisun Slough. There is much more topographic relief, however, with mountainous ter-rain dropping steeply into the river and leaving little area for marsh plain devel-opment (see figure 1.1). There are approximately 40 km² (10,000 acres) of marsh here, compared with over 260 km² (65,000 acres) north of the river.

As seen in early maps (Bache 1872; USGS 1896, 1908), most historical sloughs in the region were highly sinuous, yet they were shorter and smaller than sloughs on the north (Suisun Marsh) side of the river. Hydrodynamics were less complex, with abiotic conditions being driven strongly by events in the adjacent Sacramento–San Joaquin river channel, with some seasonal freshening pro-vided by tributaries. The marsh plain and sloughs created ecological conditions similar to those in the main Suisun Marsh on a smaller scale.

Pacheco Slough appears to have drained into large mudflats as depicted in the U.S. Coast Survey map (Bache 1872), although approximately 25 years

Peripheral Process-driven Subregions

Additional process-driven subregions in Suisun Marsh include the marsh–upland transition zone, ephemeral edge wetlands, and tidal mudflats. Although they occupy less space than subregions described above, each contributes in an important way to ecosystem functions and biotic diversity of the Marsh. These wetlands and aquatic sites form along the periphery of the marsh plain and offer key habitat for many species.

Marsh–Upland Transition Zone.   This zone borders the marsh plain where soft, waterlogged peat soils give way to firm, dry ground. This wetland type is nota-bly species rich, particularly in vascular plants. Soil composition, water content, inundation regime, and salinity change rapidly over the short distance of tran-

later the same area is depicted by USGS as a mostly vegetated emergent wetland. Otherwise, shorelines on the two maps correspond closely, suggesting that the discrepancy may point to a rapid change in land cover. Between 1896 and the present, the shoreline around the mouth of Pacheco Slough aggraded even further into former open water, which indicates that this process is ongoing. This sedimentation process was likely driven by accretion of gold-mining sediment and may currently be driven by upstream erosion from inflowing streams. Unlike the response to aggradation on Grizzly Island, dikes were not aggressively constructed to capture and drain new land in this fringing marsh; thus, the area provides an example of progressive marsh development since 1850.

Today, uplands in Contra Costa County immediately adjacent to the fringing marsh are heavily industrialized, with development including oil refineries, power plants, aggregate mines, and military installations. Railroads also pass through the region, but duck hunting clubs and agriculture were never developed here. As a result, marsh lands have not been intensively managed, and fewer dikes, diversions, ponds, and roads have been constructed on the marsh plain. Flood control channels, water management ponds, and numerous mosquito ditches are present, but the marsh plain is less degraded overall. Low-elevation marsh, high-elevation marsh, and muted tidal marsh all are proportionally well-represented landscape components (SFEI 2012). Oil spills and pollution are ongoing concerns here, much as they are in marshes north of the river. This contrast offers one possible version of what Suisun Marsh might have become in the absence of duck clubs and persistent citizen advocacy for conservation and preservation: a fragmented industrial landscape with many pockets of marsh left to develop at their own pace.

sition, yielding a wide variety of microhabitats. Historically, this wetland zone constituted a nearly uninterrupted band that bordered the marsh plain as well as marsh islands with upland areas, such as Bradmoor Island. In the contemporary landscape, this transition zone is fragmented by roads, levees, and other development. In many locations, plant diversity has been suppressed by grazing, flooding, and farming. Where remnant marsh–upland transition zone wetlands persist, as on Rush Ranch, this landscape feature is observed to support a variety of mostly native plants and animals (Wetland and Water Resources 2011).

Ephemeral Edge Wetlands. These wetlands historically dotted the marsh–upland transition zone and nearby low-lying uplands. Vernal pools and similar wetlands on stream floodplains were present, and alkali flats may also have existed. USGS topographic maps and the 1872 Coast Survey map provide clues about these former landscape features, which have since been converted to crop land, built upon, or heavily grazed. Ephemeral edge wetlands undoubtedly harbored numerous rare species and are a landscape component that, while covering a small area, supported a disproportionate amount of biodiversity (see chapter 4).

Tidal Mudflats.   Tidal mudflats were present along the northeastern edge of Grizzly Bay and at the bayward tips of Joice and Morrow islands. As recognized worldwide, mudflats are critical foraging habitat for shorebirds and fishes, and, following similar worldwide trends, the extent of mudflats has been greatly reduced in the San Francisco Estuary by human modification. Their extent presumably was affected by Gold Rush sediment even in the earliest well-surveyed cartographic depictions. During the 1900s, areas that had been mudflats in the late 1800s were largely diked or otherwise converted to other uses or wetland types. In some cases, mudflats were colonized by plants and transitioned to marsh plain or fringing marsh as a result of net sediment deposition. Approximately 2.8 km^2 (700 acres) of tidal mudflats remain in Suisun Marsh (SFEI 2012), 17% of the 16.8 km^2 (4,100 acres) surveyed in 1866–1867 (Bache 1872).

HISTORICAL HABITAT SUMMARY

Numerous processes shaped the wetland landscape that early settlers encountered in Suisun Marsh, creating a mosaic of marsh plain, sloughs, ponds, uplands, and transitional zones. Spatial and temporal variations in water quality supported a highly diverse flora and fauna with adaptations for living in a variable environment. Some of these physical and biotic characteristics have persisted into the present. Suisun Marsh is still a variably brackish system with diverse and abundant species, although many today are alien species. Other aspects of the physical and biotic environment have changed substantially since 1850.

Most of today's landscape features are easily recognized in even the oldest maps, as tides, river outflows, and wind continue to dominate physical processes. Yet hydrology is substantially different than it was 150 years ago (see chapter 3). Several moderately large sloughs (e.g., Roaring River, Grizzly, Frost, Island, upper Tree, Hastings, and Volanti sloughs) have been dammed off with dikes, affecting tidal action and connectivity for aquatic species. In channels that remain connected, circulation patterns have been altered by channel straightening, creation of new connections between sloughs, and construction of agricultural canals, mosquito-control ditches, and salinity control structures.

Human intervention since 1850 has dramatically altered distribution, quality, and quantity of different habitats and has favored some species, especially ducks, over others. Peripheral habitats, such as the marsh–upland transition zone, ephemeral edge wetlands, and tidal mudflats, have been most severely affected, to the point that it is difficult to locate examples of the first two in the modern landscape. Undiked tidal marsh has been reduced dramatically as well, with only about 31 km^2 (7,700 acres) of the 225 km^2 (55,600 acres) present in the late 1800s remaining (SFEI 2012). The balance is diked and managed, largely preventing or reversing soil-building processes. Shallow open-water environments attractive to waterfowl are much more extensive than they were historically, and they now exist throughout Suisun Marsh instead of being concentrated in the western portion. Overall, physical conditions documented in the mid-1800s through 1910 are significantly different from conditions today. In the following section, human–landscape interactions, including those that created modern conditions, are considered in more depth.

Landscape Interactions: Native Peoples

"Suisun" is the name of a southern Patwin people who lived in the vicinity of the Marsh. This was reportedly the tribe's chosen name[3] and was the moniker used by early Spanish settlers. It is rumored to translate to "land of the west wind," although it is no longer possible to verify the accuracy of the translation (Arnold 1996). Native residents throughout the region subsisted on seeds, nuts, fruits, bulbs, birds, fish, shellfish, game, and insects (Lightfoot and Parrish 2009). The record of specific human–landscape interactions is sparse, but there are enough clues to establish a sense of life on the Marsh before European contact.

The Marsh was clearly a valuable resource for food and materials. The San Francisco Bay region had likely been inhabited for 10,000 to 15,000 years (Erlandson et al. 2007) and was well populated in 1769. Some evidence suggests that native Californian peoples experienced population crashes in the 250 years prior to 1769 because the arrival of European diseases preceded explorers. Fluctuations in populations of native peoples would have had direct ecological consequences, causing plant and animal assemblages to vary in tandem (Rosenthal et al. 2007).

The area occupied by any one San Francisco Bay Area tribe at the time of European contact was usually small, estimated to be about 26 km^2 (6,400 acres). Villages typically comprised 40 to 100 people, though some were substantially larger. Population density depended somewhat on the carrying capacity of the surrounding landscape. José de Cañizares noted a settlement of about 400 inhabitants in the Carquinez Strait area in 1775, which suggests that food resources

---

3. We recognize, following Lightfoot and Parrish (2009), that the complex social structure of California Indian groups does not fit well under the conventional rubric of "tribes," which implies a fairly rigid and hierarchical structure. We nevertheless label the groups as tribes for convenience and to follow conventional usage.

of river and marsh environments were abundant (Eldredge 1909). Population reconstructions for the San Francisco region point to denser populations in areas bordering marshes because they are exceptionally productive environments (Milliken 1991).

The Suisuns were not the only group relying on marsh resources. Neighboring tribes, and their local districts (given with contemporary place names), included the Carquin to the west near Benicia, the Tolena to the north in Green Valley, the Malaca to the northeast near Maine Prairie, the Anizumne to the east near Rio Vista, the Ompin to the southeast near Antioch, and the Chupcan to the south near Martinez (Milliken 1991).

Native people were skilled boaters, crossing Carquinez Strait and San Francisco and San Pablo bays in small tule vessels. Suisun-area tribesmen regularly traversed the sloughs for both hunting and transport, as many accounts note the prevalence of tule rafts and canoes and the dexterity with which they were maneuvered (Eldredge 1909; Font [1776] 1931; Von Langsdorff 1814). One author specifically notes native hunting parties visiting Grizzly Island by boat in the mid-1800s (Duflot de Mofras 2004).

Tules (California bulrush [*Schoenoplectus californicus*] and hardstem bulrush [*S. acutus* var. *occidentalis*]) were a multipurpose construction material for people living anywhere near a tule marsh. In addition to boats, tules were used to make housing and storage structures, furniture, clothing, shoes, canoes, rafts, baskets, duck decoys, boat launches, mats, and other items. Harvesting of tules promoted new growth from rhizomes and likely increased habitat diversity (Anderson 2005). Considering that population density was generally high around marshes and that many of these items were replaced or repaired annually or more often, cutting of tules for these numerous everyday objects would have affected habitat quality in substantial areas of marsh (see chapter 4).

It is likely that Suisun-area tribes used fire as a landscape management tool in uplands and in the Marsh to flush animals and improve visibility of game (Lewis 1993). Active fires and fire scars were noted throughout California by early European explorers (Crespi 1770; Font [1776] 1931; Menzie and Eastwood 1924); however, the purposes, frequency, and spatial extent of burns in marshes are not well documented (Whipple et al. 2012). In uplands, burning clears brushy growth and encourages an oak savanna ecotype that makes harvesting plant resources and game more efficient (Lewis 1993; Anderson 2005). In marshes, burning similarly resets successional trends, and, as plants sequentially reestablish themselves, they provide habitat for a wide array of desirable species (Hackney and de la Cruz 1981). In the absence of fire, senesced vegetation becomes densely matted, excluding species that require more open habitat (Anderson 2005).

Examples of native hunting and fishing methods in the region include pitfalls for large mammals (Camp and Yount 1923), snares for smaller mammals, deerskin camouflage for deer (Von Langsdorff 1814), and nets and weirs for catching large fish such as salmon and sturgeon (Font [1776] 1931; Duflot de Mofras 2004). Nets also were used to capture birds (Maloney and Work 1943).

Waterfowl were prevalent in the diet of people living near Suisun Marsh and all along the shores of the San Francisco Estuary. Preserved whole birds stuffed with grass were offered to European visitors as a gesture of welcome or in trade (Treutlein 1972). Feathers and bird skins were also incorporated in clothing and ornamentation (Crespi 1770; Von Langsdorff 1814; Maloney and Work 1943).

People have interacted with and manipulated Suisun Marsh in ecologically significant ways more or less continually, for its entire existence. Although records are sparse, it is evident that landscape management by native peoples 250 years ago differed from management today in intent, methods, and outcomes. Species abundance and habitat mosaics in both uplands and the Marsh were the result, in part, of thousands of years of human–landscape interaction. Procuring food and shelter from the landscape for (at least) several hundred people living on the shores of the Marsh put constant pressure on natural systems. This pressure surely fluctuated along with human population, although it is not known if the number of people living around the Marsh in 1769 was relatively high or low. Hunting, fishing, foraging, and active landscape management shaped the setting that early European explorers and trappers encountered in many ways, although Europeans did not always understand what they were seeing.

Explorers, Missions, and Fur Trappers: Spanish California 1769–1822

The earliest European explorers were focused on documenting routes for travel and assessing lands for settlement; thus, any notes relevant to ecology of the region are typically motivated by one or both of these factors. Occasionally, journal entries were made on events that were outstanding in the writer's experience, such as encountering large numbers of bears or being shaken by violent earthquakes. Journal accounts affirm many conditions we would expect European explorers to have found and also present novel information about the landscape. Early exploration near Suisun was carried out on foot along the south bank of the Sacramento River and by boat (see map 4 in color insert).

In 1772, members of the Fages party were the first Europeans to view Suisun Marsh. By land, they explored the Marsh by following its southern shore to the edge of the Delta east of Antioch before heading back toward Monterey. Traveling during early spring, Pedro Fages (in Treutlein 1972) described grassy hills with wildflowers, bears, deer, elk, and antelope; he saw smoke rising from numerous villages in the area. Looking down at Grizzly Bay, he noted, "we could see that the arm of the estuary, before it made its entry between the hills, forms a sort of large, round bay, wherein were seen two or three whale calves." Fages also observed marsh lowlands and several islands with "good friable soil which can easily be drained of water" around the outlet of the Sacramento and San Joaquin rivers.

In August 1775, Don José de Cañizares, a member of Don Juan Manuel de Ayala's crew, traveled by boat through Carquinez Strait and upstream to a location near Sherman Island. Regarding San Pablo Bay, he states, "It is not difficult to enter this bay, but going out will be difficult on account of the wind from the southwest" (Cañizares 1781). This is not a surprising observation, intended as a guide to sailors that might follow him, but prescient nonetheless. His account of Carquinez Strait mentions natives who "presented us with exquisite fishes (amongst them salmon), seeds, and pinole."

In the same report, Suisun Bay is portrayed as a less promising port than Southampton Bay because "it would be difficult to obtain wood, which is far from the shore. All of the eastern coast is covered with trees; that to the west is arid, dry, full of grasshoppers, and impossible of settlement." The description of the Sulphur Springs Hills as parched and grasshopper ridden in August is in keeping with conditions today, but the account of a tree-covered eastern shoreline is unexpected. The map that accompanies the report depicts extensive stands of trees on the Montezuma Hills, on par with forests indicated in the same map around present-day San Rafael, Oakland–Hayward, and the northeastern foothills of Mount Diablo (Cañizares 1781; Eldredge 1909).

Freshwater sources and navigability of rivers were hot topics in the late 1700s. An early name for Grizzly Bay was "Puerto Dulce," or "Freshwater Port," a designation made by sailors eager to mark the transition zone between salt and fresh water, which occurred in the vicinity of Suisun Marsh (Font [1776] 1931; Eldredge 1909; Treutlein 1972). For a time, there was some debate as to whether Montezuma Slough was a substantial river of its own or merely a branch of the Sacramento River (Duflot de Mofras 2004). The Sacramento and San Joaquin rivers and the tule marshes of the Delta were major barriers to northward and eastward travel, influencing patterns of Spanish exploration and settlement in the pre-Gold Rush era.

The written record provides us with a fascinating, if narrow, vista on precolonial conditions. For example, in April 1776, at the northeastern foot of Mount Diablo, tule elk (*Cervus canadensis nannodes*) were described by Pedro Font ([1776] 1931) as follows:

> On descending to the plain we saw, near the water and about a short league away, a big herd of large deer, being, I think, what they call "buros" in New Mexico. They are about seven spans high, and have antlers about two yards long with several branches. Although an effort was made to get one, it was impossible, because they are very swift, and the more so at this time as they had shed their great antlers, which undoubtedly they do at seasons judging from the many horns that we saw lying about. All this region abounds in these deer; and the tracks, resembling those

of cattle, that we found this day and the next, make it appear as if there was an immense herd of cattle thereabouts.

Tule elk are mentioned in early accounts more frequently than pronghorn antelope (*Antilocapra americana*) or mule deer (*Odocoileus hemionus*), possibly because they were more prevalent when European explorers arrived or because they were more visually striking animals. Grizzly bear (*Ursus arctos horribilis*), beaver (*Castor canadensis*), and river otter (*Lontra canadensis*) were among other mammals in high abundance in and near Suisun Marsh around 1800. It is possible that reductions in human populations due to waves of disease meant that animal populations in the 1770s were relatively high in comparison to populations in previous decades or centuries (Preston 1998). Sizeable populations of such animals would have had meaningful landscape interactions, exerting grazing pressure on plant communities, trampling pathways through the Marsh, wallowing in standing water, dust bathing in salinas, and affecting soil processes.

Soon after the first explorers marched along the shores of San Francisco Bay, Spaniards began to settle the region, building missions, presidios, and pueblos. The first non-Spanish settlers appeared shortly thereafter, among them fur trappers and whalers. In Suisun Marsh, this period involved a major drop in human population and overexploitation of fur-bearing wetland species such as beaver and river otter.

Following the establishment of Mission San Francisco in 1776, Mission Santa Clara in 1777, and Mission San Jose in 1797, native populations of the San Francisco Bay region were assimilated over the course of several decades. The Suisuns actively resisted conversion between 1800 and 1810, becoming revered by neighboring tribes—and notorious among Europeans—for their unwillingness to cooperate[4] (Schoolcraft et al. 1857). Many individuals who left Mission San Francisco, considered runaways by the Spanish, came to Suisun to avoid recapture. For a time, the Suisun Marsh area was perceived as a borderland by both natives and colonists. Between 1810 and 1816, the Suisuns' will to resist waned, and most abandoned their villages for Mission Dolores in San Francisco. This event was disheartening for other tribes in the region. Once the formerly steadfast Suisuns ceased to resist missionization, surrounding tribes joined the missions in rapid succession (Milliken 1991).

Mission establishment marked major changes in landscape ecology. By the 1830s, the lands of the San Francisco Bay region were largely depopulated, both

4. By the mid-1800s, the relationship between Suisuns and colonists had reversed, with the Suisuns being favored by Mexicans (particularly Mariano Vallejo) for their character and abilities. One author noted that "In Petaloma [sic] valley, the original inhabitants are reduced to almost nothing, and they have been replaced by the Indians of Suisun, from the bay of that name, above Benicia" (Schoolcraft et al. 1857).

because tribes had left their villages for the missions and because measles, cholera, smallpox, dysentery, and other introduced diseases had killed an enormous number of people (Milliken 1995; Patterson and Runge 2002; Ahrens 2011). Both emigration and disease-related death were centered geographically around missions, radiating outward from them in roughly decreasing degrees of severity. Given the uneven spatial distribution of impacts, estimates of a 20% statewide native population decline for this period (Cook 1976) likely understate the much more dramatic decrease in population on the shores of the San Francisco Estuary.

In 1833, George C. Yount traveled through central California on a trapping expedition and witnessed the aftermath of a major malaria outbreak. After relaying bleak scenes, including abandoned villages, the "bones of untold thousands" lying in the valleys, and encounters with despondent, starving survivors, he reflects on the effects of nearly 40 years of human depopulation on regional ecology. As recorded by Clark (in Camp and Yount 1923):

> In 1833—Benicia was visited and has been thus described: It was then nothing more than a wide and extended lawn, exuberent [sic] in wild oats and "a place for wild beasts to lie down in"—The Deer, Antelope and noble Elk held quiet and undisturbed possession of all that wide domain, from San Pablo Bay to Sutter's Fort. . . . The above named animals were numerous beyond all parallel—In herds of many hundreds, they might be met, so tame that they would hardly move to open the way for the traveller [sic] to pass—They were seen lying, or grazing, in immense herds, on the sunny side of every hill, and their young, like lambs were frolicing [sic] in all directions—The wild geese, and every species of water fowl darkened the surface of every bay, and frith, and upon the land, in flocks of millions, they wandered in quest of insects, and cropping the wild oats which grew there in richest abundance—When disturbed, they arose to fly, the sound of their wings was like that of distant thunder—The Rivers were literally crouded [sic] with salmon, which, since the pestilence had swept away the Indians, no one disturbed—It was literally a land of plenty, and such a climate as no other land can boast of.

One point of interest in this description is that, in 1776, Font and his men found elk hiding in the tules notably shy of humans, yet Yount found emboldened herds covering nearby hills less than 60 years later. Today, imagining either scene unfolding in the hinterlands between Mare Island and midtown Sacramento requires a certain suspension of disbelief. Yount's impression was echoed by other travelers during this era, and, even if his account contains some embellishment, it points to profound changes in the landscape brought by a sudden drop in human population. In sum, large-scale environmental responses included decreased hunting pressure on animal populations, increased grazing pressure on grasslands, and decreased foraging and manipulation of plant communities by native peoples. With an increase in prey, large predators such as grizzly bears increased in abundance as well (Preston 1998).

Yount surveyed the hills around Suisun in a key period of ecological transition. In 1833, missions had substantially diminished the human population, but ranchos had yet to disperse their vast cattle herds into the area. Not a single large-scale rancho existed north of San Francisco Bay, and the nascent San Antonio, San Pablo, and Pinole ranchos, located between the East Bay Hills and San Francisco Bay shoreline, were the only ranchos east of the bay. Missions were established in San Rafael in 1817 and Sonoma in 1823, but these were relatively remote from Suisun. In about 1825, a small rancho and *asistencia,* or submission, was established near present-day Rockville. This tiny outpost was run by Christian Indians who were charged with raising crops, running cattle, and converting non-Christians from outlying tribes (Bowen 2009).

The mission era marked the onset of steady introductions of European animals such as cows, goats, pigs, horses, and sheep into the environment. Old World plant species were likely already a part of California's flora when the first explorers arrived, but abundant sustained introductions during the mission era caused alien plants to colonize prodigiously (Mensing and Byrne 1998; Preston 1998; Minnich 2008). Much like effects on human populations, these changes radiated outward from missions and ranchos.

Beavers and river otters were trapped in abundance during this period, with reports of 4,000 beavers taken from Suisun in 1830 alone (Duflot de Mofras 2004). In the spring of 1833, trapper John Work and his party camped on the periphery of the Marsh for a few weeks. His journal affirms the presence of numerous deer, antelope, elk, and bears but only a handful of native people. His party killed at least 17 bears during their stay. Work was keenly interested in assessing the abundance of beavers, and he gives us this characterization of the Marsh: "[Suisun] bay is destitute of wood, it has the resemblance of a swamp overgrown with bulrushes and intersected in almost every direction with channels of different sizes and except the want of wood apparently very well adapted for beaver, the people say that beaver are to be found among the rushes" (Maloney and Work 1943). Work did not report catching many beavers in the Suisun area, probably because large numbers were taken just a few years before. The beaver and river otter were hunted down to very low numbers in Suisun Marsh and the Delta before demand for their pelts waned, and populations have never fully recovered.

Insofar as wildness is defined as lack of human intervention, conditions in Suisun Marsh in the early 1830s were perhaps wilder than they had ever been. Game populations soared in the absence of hunting pressure and active habitat management by humans. Land cover and ecosystem functions no doubt were affected by the cease in intentional fire-setting for vegetation management and by the spread of nonnative plants. This period was followed by settlement activities that further affected regional ecology (see table 2.1).

TABLE 2.1 Early European exploration and settlement.

Year	Event	Description
1542	Territory claimed by Spain	Juan Rodriguez Cabrillo explored the Pacific coast of North America; Spain claimed the Californias.
1772	First European land expedition encountered Suisun Marsh	Traveling along the south shore of Suisun Marsh, the Fages party encountered friendly Indians harvesting spring-run salmon. They prepared a map of San Francisco Bay, San Pablo Bay, and the Suisun area.
1775	First Nautical expedition: the *San Carlos* arrived at San Francisco Bay	The *San Carlos* was the first ship to pass through the Golden Gate. Lieutenant Juan de Ayala was injured, so first pilot Don Jose Cañizares and second pilot Don Juan B. Aguirre explored and mapped Suisun Bay.
1776	Second European land expedition encountered Suisun Marsh	The Suisun south shore was traversed by Captain Juan Bautista de Anza, Father Fray Pedro Font, and their party. They attempted to trade glass beads for fish, but the Indians would only trade for clothing.
1776	San Francisco Presidio and Mission established	Construction of San Francisco Presidio and Mission Dolores began.
1806	Russians visited San Francisco	A Russian party led by Langsdorff lodged on San Francisco peninsula and traveled by boat to south San Francisco Bay. Their goals were to obtain supplies and gain intelligence on Spanish California.
1809	Carquins moved to mission at San Francisco	The Carquins abandoned villages for Mission Dolores, leaving no buffer between Suisuns and missionaries.
1810–22	Suisuns moved to mission at San Francisco	The Suisuns abandoned villages to join the mission at San Francisco. Those who did not want to go to the mission moved east to join other Patwin tribes.
1812	Russian settlement at Fort Ross founded	The Spanish perceived Russian settlement as a territorial threat, increasing their determination to maintain missions.
1817	Mission San Rafael founded	Mission San Rafael Arcangel was established as a sanitarium in 1817 and became a full-fledged mission in 1822.
1823	Mexican Republic instated	This was a precursor to the fall of missions.
1823	Altimira explored north bay and established mission in Sonoma	Father Jose Altimira searched for a mission site between Petaluma and Suisun and then founded Mission San Francisco Solano in Sonoma. This is the only mission established during Mexican rule.
1824	Santa Eulalia Asistencia established by Altimira	Jose Altimira established the Santa Eulalia Asistencia and rancho near Rockville. This is probably the first sustained Spanish influence (cattle, plants, culture, etc.) near Suisun Marsh.
1833	Yount trapping foray	George Yount traveled through the Suisun area while hunting beaver and otter.
1833	Work Expedition	John Work's expedition camped at Suisun Creek while members went to the mission at Sonoma for ammunition.
1840–46	Mexican land grants near Suisun	Cattle ranching ratcheted up; in the Suisun area, ranchos included Suisun, Tolenas, Los Ulpinos, and Chimiles, totaling over 269 km^2 (66,000 acres).

Alta California: Ranchos of the Mexican Republic 1822—1848

Between the late 1830s and 1848, the Bay–Delta region experienced the rapid proliferation of ranchos (see map 4 in color insert). Though the effects took a few years to reach the extremities of Alta California, the transition from Spanish-chartered missions, presidios, and colonies to Mexican territory in 1821 changed more than the area's sociopolitical landscape. Animals, particularly cattle, became the basis of California's economy. During the 22-year span of the Mexican era, over 6 million cow hides were shipped from California (Hackel 1998). Vast, essentially wild herds of cattle put increasing pressure on native plant populations and eventually displaced native grazers (Preston 1998).

Mexican-style ranchos were short lived in the Suisun area, because all were granted in the decade prior to the Gold Rush (see map 4). This was enough time, however, for even a small initial herd of cattle to multiply into the thousands (Von Langsdorff 1814; Menzie and Eastwood 1924). Ranchos in the region were not located in the Marsh proper, although it was customary to allow cattle to roam freely and use any available freshwater sources as watering holes.

Diseños and Land Case maps drawn for legal purposes in the 1840s show the beginnings of hydrologic management in areas adjacent to Suisun Marsh. In these years, the first water diversions were constructed to irrigate small plots of vegetables and grain.

Trapping of fur-bearing animals continued during the Mexican era but tapered off gradually. The Mexican government made business increasingly difficult for foreigners, and overtrapping had reduced populations of the most desirable species, such as beaver and river otter, to unprofitable levels by the 1840s. At the same time, market demand for pelts was dropping. The Russians abandoned Fort Ross in 1841, and Hudson's Bay Company closed their San Francisco office in 1846 (Ogden 1933; Maloney 1936). Resident Californians continued trapping at reduced levels, however, keeping populations of target species low.

The human population and human–landscape interactions shifted fundamentally in this time frame, moving from hundreds of people engaged in active landscape-scale management for preferred native species to a handful of residents farming cattle and vegetables, and then to large-scale rancho development. As of 1848, the Marsh plain and sloughs were still functionally intact, with only minor modifications for irrigating small-scale agricultural plots in adjacent uplands.

Early American California: 1848–1900

Between 1848 and 1850, California became a U.S. territory, underwent explosive growth with the Gold Rush, and was granted statehood. Hundreds of thousands of people poured into the San Francisco region, and the pace and extent of landscape-level change mushroomed. Suisun Marsh was not at the center of these changes, but it was on the immediate periphery. Sustained intensive waterfowl hunting, farming, settlement, land speculation, industrialization, and militarization all came to the Marsh in this era, at which time it began to look and function like the Marsh we are familiar with today.

Those who arrived early on found a landscape much like the one described by Yount in 1833. After 1849, however, fauna rapidly declined, both in individual abundances and in population distributions: animals including birds, elk, antelope, deer, and bears were killed for subsistence, sport, and sale at market. Again the spatial effects of settlement radiated outward, this time from both San Francisco and Sacramento. In 1855, the following descriptions were made (Newberry 1857):

> In the rich pasture lands of the San Joaquin and Sacramento, the old residents tell us, it [tule elk] formerly was to be seen in immense droves, and with the antelope, the black-tailed deer, the wild cattle, and mustangs, covered those plains with herds rivalling [sic] those of the bison east of the mountains, or of the antelope in south Africa.

> Though found in nearly all parts of the territory of the United States west of the Mississippi, it [antelope] is probably most numerous in the valley of the San Joaquin, California. There it is found in herds literally of thousands; and though much reduced in numbers by the war which is incessantly and remorselessly waged upon it, it is still so common that its flesh is cheaper and more abundant in the markets of the Californian cities than that of any other animal. . . . In the Sacramento valley they have become rare, and the few still remaining are excessively wild.

Hunting of Suisun Marsh waterfowl for the San Francisco market began around 1859. Market hunters were opportunists, profiting by harvesting abundant waterfowl from unregulated wildlands to feed the growing human population in San Francisco (Arnold 1996). They were not systematic in recording their kills, so their accounts cannot be used as an indication of relative species abundance. Generally, market hunters' descriptions of bird populations continue in the same vein as descriptive accounts prior to 1849, further corroborating the productivity of the region.

With the Swamp Land Act of 1850, Congress granted "swamp land"—in other words, marshes—to the State of California, thus facilitating reclamation of wetlands. This legislation instigated a rush to patent, drain, dike, and cultivate marshes. Marshes nearer to population centers were claimed first; thus, Suisun Marsh initially was not a target for reclamation and land development. Settlement did come eventually, however, as some farmers sought to recreate the wild success of early agriculture in the Delta.

Lands were patented, surveyed, parceled out, and sold beginning in the late 1800s. Farms appeared throughout the southeastern Marsh, at the western foot of the Potrero Hills and in the northeastern Nurse–Denverton slough region. The era of reclamation in Suisun was less comprehensive than in the Delta, but its effects are still evident in today's landscape. By the 1880s, reclamation districts had been organized and dike and levee construction to protect agriculture was well underway. Most construction was completed before 1920; by 1930, 181 km^2 (44,600 acres) had been enclosed (Miller et al. 1975).

During this same period, the marsh plain was increasing in size because of the deposition of sediment washed down from hydraulic gold mining in the Sierra Nevada, allowing enterprising landowners to dike off new shoreline. Between 1900 and 1953, Grizzly Island reclamation efforts increased the size of the island by nearly 20 km^2 (5,000 acres), or 40% (map 5 in color insert). Today, the southwestern portion of Grizzly Island supports annual and perennial grasses and is primarily used for tule elk pasture.

Suisun Marsh's location approximately midway between San Francisco and Sacramento made it a convenient place for ships to take on agricultural products for urban markets in the mid- to late 1800s, when waterways were primary shipping routes. For a time, agriculture in Suisun Marsh followed the same trajectory as in the Delta. Settlers constructed boat landings at numerous locations that enabled export of dairy, cattle, and agricultural products. Building infrastructure to control tides and flooding in wetlands was labor intensive but could potentially pay off when nutrient-rich marsh soils produced high crop yields. As in the Sacramento–San Joaquin Delta, water for irrigation was abundant. Notably, water from wells and diversions in Suisun through the early 20th century was consistently fresh enough for irrigation of berries, fruit trees, wine grapes, oats, barley, corn, beans, hay, and asparagus, and for the watering of cattle and pigs. Additionally, tules and California cordgrass (*Spartina foliosa*), for use as packing material, and clay, for use in the ceramic industry, were exported from the Marsh (Frost 1978).

Crop yields from the early years of farming were good, but, beginning in the late 1920s, exterior economic pressures and increasing salinity driven by droughts and upstream water diversions led to the eventual decline and abandonment of farms. There was a short-lived move to dairying, but today agricultural activities are confined to upland areas adjacent to the Marsh. Frost (1978) noted that, in 1927, "[The Baby Beef Company's] willingness to buy land on [Grizzly] Island, combined with the depression and Shasta Dam cutting down the fresh water supply, began a decline in dairying on Grizzly Island. The idea of owning duck hunting clubs was also gaining in popularity." The west side of the Marsh was also patented, and some efforts toward reclamation were undertaken, but farming never took hold there as it did in the southeastern Marsh. The geography of the western Marsh, with its numerous natural ponds, attracted both ducks and duck hunters.

Duck hunting and railroad construction went hand in hand. As wealthy San Franciscans looked to the Marsh for recreation, they found that quick transportation was essential. A railroad line running directly through the Marsh—and to the doorsteps of the best clubs—was rumored to be the most expensive track, mile for mile, in Southern Pacific's history because substantial stretches repeatedly sank into the soft peaty soils (Arnold 1996). As reported 34 years after construction began, "Southern Pacific trains resumed travel today over the section of the Suisun [M]arsh, which swallowed up the track Friday. The marsh had sunk about twelve feet, for a stretch of more than fifty feet. The Suisun [M]arsh has presented one of the most puzzling problems with which the officials of the division of the road have had to deal. The 3000 feet of track across it has frequently sunk and many attempts to build a firm foundation for the tracks have failed" (*Los Angeles Times,* June 5, 1912).

Railroad construction spurred land speculation in adjacent areas, yet inflated property values appeared to help Suisun Marsh, unlike many nearby marshes, evade development. Sometimes threats were avoided by luck, sometimes through citizen action, and sometimes because the people proposing changes were merely charlatans intending to defraud unwary investors (table 2.2). An increased awareness of the socioeconomic value of wetlands finally led to preservation agreements that have protected the Marsh for the past 40 years.

20th Century: Advent of the Conservation Ethic

Duck clubs ruled the Marsh throughout the 20th century, and members' persistent advocacy thwarted numerous campaigns to industrialize the area (see table 2.2). Duck club management efforts have resulted in expanded habitat for ducks, and in a far "softer" landscape (as opposed to an urbanized, "rigid" landscape) than in developed Delta and San Francisco Bay wetlands. The wetland habitat intended primarily for waterfowl supports a broad array of species and lends itself much better to marsh restoration. At the same time, duck clubs have engineered the landscape for their own purposes, altering ecological processes significantly. Islands have been diked and ponds managed to favor vascular plants preferred by sport waterfowl (Arnold 1996). Management legacies include subsidence of peat soils due to pond leaching cycles and occasional but recurrent fish kills when anoxic pond water is released into adjacent sloughs. Increasing pressures brought by sea-level rise will exacerbate existing problems.

While efforts have been made to stabilize hydrologic conditions in Suisun Marsh and maximize its potential to support migrating waterfowl, landscape change has occurred on all sides. Remarkably, the Marsh managed to avoid the sweeping conversion of wetlands seen elsewhere in the San Francisco Estuary. Increased awareness of environmental issues by the general public has buoyed

TABLE 2.2 Threats to Suisun Marsh and nearby areas over 125 years.

Year	Event	Description
1846	Montezuma City	Montezuma City, intended to be a bustling Mormon colony, founded by Lansing W. Hastings. The plan never came to fruition.
1847	City of Francisca (Benicia) founded	A land speculator named Robert Semply planned a sprawling city in the present-day location of Benicia.
1848	Suisun City founded	Started as a bustling center for shipping via water and rail, but economy faltered after city was bypassed by highway I-80 in 1963.
1850	New York of the Pacific (Pittsburg) founded	Colonel Jonathan D. Stevenson founded "New York of the Pacific" at the present-day site of Pittsburg, hoping for a booming metropolis to rival New York City.
1853–54	Benicia declared capital of California	Benicia held sway as California's capitol for nearly 13 months. It would have become a much larger city if the capitol had not been moved to Sacramento.
1860s	Earliest salinity control planning	First discussions about shutting tide water out of Suisun Marsh to control salinity occur.
1860s	Collinsville (Newport City) founded	Three successive owners plan large town at present-day Collinsville. The town perpetually failed to thrive.
1880	Carquinez Strait tidal barrier	State Engineer W. W. Hall proposed a tidal barrier across Carquinez Strait to prevent saltwater intrusion.
1912–13	Calhoun's Solano City swindle	Patrick Calhoun invested over $1 million of public utility money in publicity and creation of a navigable channel to "Solano City" and "Solano Irrigated Farms" at the upstream end of Lindsey Slough. He later transferred ownership of worthless shares in "Solano Irrigated Farms" to the utility to compensate for the money he took.
1921; 1924	Dam from Richmond to San Quentin	Captain C. S. Jarvis of the U.S. Army Corps of Engineers (USACE) proposed a massive dam-and-lock project to turn San Pablo Bay and upstream waters into a giant freshwater reservoir.
1946	Reber Plan	John Reber presented the "ultimate solution" to Bay–Delta management "problems." His plan included a dam from San Quentin to Richmond, a 2,000-ft-wide causeway south of the Bay Bridge, ship locks, and a giant dredged ship channel along the east bay shoreline. A $2.5 million USACE feasibility study was undertaken. Drawbacks were determined to outweigh benefits. Two lasting vestiges exist: the Bay circulation model in Sausalito and the Bay Conservation and Development Commission.
1974	Giant garbage dump	Proposal was made to barge large quantities of garbage from urban areas to Potrero Hills, which would have involved major dredging and channel straightening in the Marsh. Present-day operation uses trucks.
1975	Dow Chemical Plant	Dow Chemical proposed $500-million project in Collinsville. Construction was fought down by concerned citizens and would have required 65 government permits; a sign of changing attitudes about the environment.

SOURCES: *Los Angeles Times*, March 10, 1974; *New York Times*, January 21, 1977; Hogan and Papineau 1980; Arnold 1996; Stone 1996.

TABLE 2.3 Environmental legislation, actions, and agreements affecting Suisun Marsh.

Year	Event
1963	Suisun Resource Conservation District formed
1965	San Francisco Bay Conservation and Development Commission formed
1970	Four-Agency Memorandum of Agreement
1972	Clean Water Act
1973	Endangered Species Act
1974	Nejedly-Bagley-Z'Berg Suisun Marsh Protection Act
1976	Suisun Marsh Protection Plan
1977	AB 1717: The Suisun Marsh Protection Act of 1977
1978	State Water Resources Control Board (SWRCB) Water Rights Decision 1485
1979–80	Roaring River and Morrow Island distribution systems and Goodyear Slough outfall constructed
1984	Plan of Protection for Suisun Marsh
1987	Suisun Marsh Preservation Agreement (SMPA)
1988	Suisun Marsh Salinity Control Gate constructed
1990–95	Planning for the Western Salinity Control Project
1991	Cygnus and Lower Joyce facilities constructed
1994	Bay–Delta Accord
1995–98	SWRCB Water Quality Control Plan
1995	Amendment 3 to the SMPA
1999	SWRCB Water Rights Decision 1641
2000	Draft Jeopardy Biological Opinion
2000	CALFED Suisun Marsh Charter
2001	Suisun Marsh Charter Implementation Plan
2001	CALFED Grant Proposal
2003	Habitat Management, Preservation, and Restoration Plan
2004	Bay Delta Science Consortium Suisun Marsh Science Workshop
2011	Suisun Marsh Habitat, Preservation, and Restoration Plan

legal protection for the Marsh (see table 2.3). Compared to the Delta, Suisun Marsh lacks extensive reinforced slough and channel banks, heavily fortified levees, extreme hydrologic manipulation, and largely agrarian land use. Compared to wetlands on the shores of San Pablo and San Francisco bays, it lacks major dredge-and-fill projects, shipping ports, and industrial sites. These landscapes are civil-engineered endpoints, hardened in ways that make them poor candidates for wetland ecosystem restoration. Thus, the presence of less-altered wetland ecosystems is one reason developers look to Suisun Marsh for mitigation sites when wetlands are eliminated elsewhere. Suisun Marsh's relative importance as a wildlife refuge increases with every nearby parcel of wetland converted to another use.

Construction of hydrologic infrastructure upstream of Suisun Marsh, including the Central Valley Project and the State Water Project, has had serious effects on the quantity and quality of water entering the Marsh. In the mid-1970s, civil plans to increase diversions in the Sacramento–San Joaquin Delta were expected to raise salinity in the Marsh, which was perceived as a threat to water-quality conditions required for the maintenance of waterfowl populations. The Suisun Marsh Plan of Protection and Water Rights Decision 1485 addressed this threat with yet more engineering, clearing the way for construction of major water-distribution facilities in the Marsh. The salinity control gates and Roaring River and Morrow Island distribution systems, which increase the net inflow of fresh water, were mandated by this legislation (Sweeny and Spencer 1984).

Entering an Era of Multispecies Management

In 1976, the San Francisco Bay Conservation and Development Commission established the Suisun Marsh Plan of Protection, a mitigation measure meant to compensate for impacts of major water projects on California's Central Valley. This was the beginning of a new management approach wherein government agencies began seriously investing in the Marsh's mitigation potential. Most efforts have aimed to bolster waterfowl populations, but preserves have also been established to encourage locally restricted species such as Suisun thistle (*Cirsium hydrophilum* var. *hydrophilum*), salt marsh harvest mouse (*Reithrodontomys raviventris*), and Suisun shrew (*Sorex ornatus sinuosus*). With endangered-species awareness comes field science, and the Marsh has been the site for hundreds of natural history studies, covering topics as diverse as sediment cores and jellyfish invasions.

KEY IMPLICATIONS OF SUISUN MARSH HISTORICAL ECOLOGY

· For its entire 6,000-year history, Suisun Marsh has experienced both wetter and drier episodes lasting a few years to hundreds of years. This inherent regional variability is likely to increase when combined with the predicted effects of climate change. The fairly benign, largely freshwater conditions experienced in the past 150 years appear to be an anomaly.
· Suisun Marsh is a complex landscape. It has never been simply one monotonous tule wetland. Many medium-scale habitats (ponds, salinity gradients, sloughs, marsh plain, upland transition zone, edge wetlands) exist, generated in great part by geomorphic diversity. This habitat diversity has favored a high richness and abundance of fish, wildlife, and vascular plants.
· Even when the earliest explorers surveyed the islands and waters of Suisun Marsh, this vast wetland mosaic was in flux. There has never been long-term stasis in the Marsh at geologic or even human time scales. It is a landscape that responds quickly to human-induced change, an intrinsic quality that can be leveraged for ecosystem restoration.

- There is no such thing as a "premanagement" Marsh. Native peoples had strong interactions with the landscape, modifying it in many ways, for example by pruning, harvesting, and burning plants, reducing wildlife populations through hunting, and modifying animal behavior and distribution through resource manipulation. Their management, in tandem with landscape interactions with beavers, elk, waterfowl, and other vertebrates, likely resulted in a tidal marsh characterized by moderately sized expanses of open water and patches of tules in different stages of succession at all times.
- The history of Suisun Marsh suggests that management of duck clubs and wildlife areas will eventually have to change, even in the absence of climate change. Freshwater inflow is likely to decline and/or become more variable, while subsidence of nontidal lands and pollution of Marsh sloughs will need to be halted (see chapter 3). Such a regime shift in management is historically compatible with past major changes in Marsh management.
- Connectivity to outlying habitats was, and continues to be, critical in allowing Suisun Marsh to meet the needs of native migratory mammals, birds, and fishes. Plans to use Suisun Marsh wetlands to compensate for wetland or protected species loss elsewhere through mitigation and habitat restoration projects will need to take into account the importance of interconnectedness with the larger landscape.

IN SUMMARY:
WAS THE HISTORICAL MARSH DIFFERENT?

Compared to the nearby Delta, Suisun Marsh *looks* more natural and is often perceived as a "green" space minimally altered by humans. The lack of conspicuous urban development tends to mislead casual observers into thinking that the entire marsh area is managed public land existing in a natural state. In fact, less than 20% of the Marsh is public land. Remaining areas are privately owned, primarily by duck hunting clubs. Conservation on both private and public land in Suisun Marsh most often means managing and/or creating habitat to attract waterfowl, with a few publicly owned reserves managed for special-status species.

Suisun Marsh was very different before the introduction of hydrologic infrastructure and modern landscape management. Salinity levels in bayward portions of the Marsh varied more than they do today. In the absence of reinforced dikes, major precipitation events caused widespread flooding of the marsh plain. Although sea level was rising during most of the early history of the Marsh, soil was accreting at a roughly commensurate rate. The configuration and proportion of habitats such as ponds, marsh plain, and mudflats were significantly different than they are now. When the Marsh was functionally intact as a tidal marsh system, so were other ecosystems throughout the western Americas.

Interrelationships between the Marsh and these places allowed it to support immense numbers of migratory and resident animals, as well as a large human population.

Beginning in the late 1800s, hydrologic infrastructure, created to keep tidal water from encroaching on farmlands and duck clubs, significantly modified the appearance and functions of the marsh plain and sloughs. The addition of large quantities of sediment from hydraulic mining further affected the shape of the Marsh. Many natural landscape features and processes have been obscured by dikes, upstream dams and diversions, constructed wetlands, salinity control structures, roads, and railroads. The cumulative effects of the introduction of approximately 7,000 dams and diversions upstream in the Sacramento and San Joaquin river watersheds (CALFED 1996), in particular, have dramatically altered the timing and volume of freshwater inflows. Subsidence as the result of wetland management for duck hunting is a problem in Suisun Marsh, although it is often overlooked because loss of elevation is not as extreme as in the Delta (chapter 3) and because management objectives are different. Today, about 90% of the marsh plain has been diked (California Department of Water Resources 2010). This prevents tidal inundation and, thus, sediment delivery, except in the case of dikes overtopping during major floods or exceptionally high tides. Pond management can leave peaty soils exposed to atmospheric oxygen, causing them to decompose and become compacted. Under these conditions, soil-building processes slow down or are reversed. In concert, these physical changes have had major impacts on landscape form, sediment delivery, hydrologic circulation, and soil formation in the Marsh.

Understanding how Suisun Marsh has changed with climatic variability and human management over the past 6,000 years provides insights on how to guide change in the future. Increasing hydrologic connectivity to counteract subsidence has a much better chance of succeeding in Suisun Marsh than in much of the Delta, which has far greater levels of subsidence with which to contend. The Marsh will continue to change and will likely play an increasingly important role in the conservation of the biota of the San Francisco Estuary and surrounding region.

REFERENCES

Ahrens, P. 2011. John Work, J. J. Warner, and the Native American Catastrophe of 1833. Southern California Quarterly 93(1): 1–32.

Anderson, M. K. 2005. Tending the Wild: Native American Knowledge and the Management of California's Natural Resources. Berkeley: University of California Press.

Arnold, A. 1996. Suisun Marsh History: Hunting and Saving a Wetland. Marina, California: Monterey Pacific.

Atwater, B. F., S. G. Conard, J. N. Dowden, C. W. Hedel, R. L. MacDonald, and W. Savage.

1979. History, landforms, and vegetation of the Estuary's tidal marshes. Pages 347–385 *in* San Francisco Bay: The Urbanized Estuary (T. J. Conomos, A. E. Leviton, and M. Berson, eds.). San Francisco: AAAS Pacific Division.

Atwater, B. F., C. W. Hedel, and E. J. Helley. 1977. Late Quaternary depositional history, Holocene sea-level changes, and vertical crust movement, southern San Francisco Bay, California. Geological Survey Professional Paper 1014. Washington D.C.: U.S. Geological Survey.

AWS Truepower. 2010. California—Annual Average Windspeed at 80m. Albany, New York: AWS Truepower.

Bache, A. D. 1872. Chart Number 626 Suisun Bay California. 1:40000. U.S. Coast Survey. National Oceanic and Atmospheric Association. http://historicalcharts.noaa.gov/historicals/.

Balling, S. S., and V. H. Resh. 1983. The influence of mosquito control recirculation ditches on plant biomass production and composition in 2 San Francisco Bay USA salt marshes. Estuarine Coastal and Shelf Science 16: 151–162.

Bowen, J. 2009. Old stone building sparks new historical research and discoveries. Solano Historical Society, Fairfield, California.

Burroughs, W. J. 2007. Climate Change: A Multidisciplinary Approach, 2nd ed. New York: Cambridge University Press.

CALFED. 1996. CALFED Bay–Delta Program Phase I Final Report. CALFED Bay–Delta Program.

California Department of Water Resources. 2010. Suisun Marsh Program. California Department of Water Resources. http://www.water.ca.gov/suisun/.

Camp, C. L., and G. C. Yount. 1923. The Chronicles of George C. Yount: California Pioneer of 1826. California Historical Society Quarterly 2: 3–66.

Cañizares, J. de. 1781. Plan del gran Puerto de San Francisco [Map of the Grand Port of San Francisco Bay]. In the collection of the Bancroft Library, University of California, Berkeley.

Collins, J. N., and R. M. Grossinger. 2004. Synthesis of scientific knowledge concerning estuarine landscapes and related habitats of the South Bay Ecosystem. Technical report of the South Bay Salt Pond Restoration Project, Oakland, California.

Cook, S. F. 1976. The Population of the California Indians 1769–1970. Berkeley: University of California Press.

Crespi, J. 1770. A Description of Distant Roads: Original Journals of the First Expedition into California, 1769–1770. San Diego, California: San Diego State University Press.

Drexler, J., C. de Fontaine, and T. Brown. 2009. Peat Accretion Histories During the Past 6,000 Years in Marshes of the Sacramento–San Joaquin Delta, CA, USA. Estuaries and Coasts 32: 871–892.

Duflot de Mofras, E. D. 2004. Travels on the Pacific Coast: A Report from California, Oregon, and Alaska in 1841. Santa Barbara, California: Narrative Press.

Eldredge, Z. S. 1909. The March of Portola and the Discovery of the Bay of San Francisco. San Francisco: California Promotion Committee.

Erlandson, J. M., T. C. Rick, T. L. Jones, and J. F. Porcasi. 2007. One if by land, two if by sea: who were the first Californians? Pages 394 *in* California Prehistory: Colonization, Culture, and Complexity (T. L. Jones and K. A. Klar, eds.). Lanham, Maryland: Altamira Press.

Font, P. (1776) 1931. A chronicle of the founding of San Francisco. *In* Font's Complete Diary: A chronicle of the founding of San Francisco. Translated and edited by H. E. Bolton. Berkeley: University of California Press.

Frost, J. 1978. A Pictorial History of Grizzly Island. San Francisco: Trade Pressroom.

Hackel, S. W. 1998. Land, labor, and production: the colonial economy of Spanish and Mexican California. Pages 111–146 *in* Book Land, Labor, and Production: The Colonial Economy of Spanish and Mexican California (R. A. Gutierrez and R. J. Orsi, eds.). Berkeley: University of California Press.

Hackney, C. T., and A. A. de la Cruz. 1981. Effects of fire management on brackish marsh communities: Management Implications. Wetlands 1: 75–86.

Hall, M. 2004. Comparing channel form of restored tidal marshes to ancient marshes of the north San Francisco Bay. Hydrology. Water Resources Center, University of California, Berkeley.

Hogan, C. M., and M. Papineau. 1980. Air Quality Impact Analysis of the Proposed Dow Collinsville Plant. Earth Metrics Incorporated Solano County.

Jepson, W. L. 1905. Where ducks dine. Sunset 14: 409–411.

Lacy, J. R., and S. G. Monismith. 2000. Wind, sea level, and a sudden increase in salinity in northern San Francisco Bay. Washington, D.C.: American Geophysical Union.

Lewis, H. T. 1993. Patterns of Indian burning in California: ecology and ethnohistory. Pages 55–116 *in* Before the Wilderness: Environmental Management by Native Californians (T. C. Blackburn and M. K. Anderson, eds.). Menlo Park, California: Ballena Press.

Lightfoot, K., and O. Parrish. 2009. California Indians and Their Environment: An Introduction. Berkeley: University of California Press.

Loeb, D. 2011. A Life in Geologic Time: Learning the Landscape with Doris Sloan. Berkeley: Bay Nature.

Malamud-Roam, F., D. Dettinger, B. L. Ingram, M. K. Hughes, and J. L. Florsheim. 2007. Holocene climates and connections between the San Francisco Bay Estuary and its watershed: a review. San Francisco Estuary and Watershed Science 5(1): article 3.

Malamud-Roam, F., and B. L. Ingram. 2004. Late Holocene $\delta^{13}C$ and pollen records of paleosalinity from tidal marshes in the San Francisco Bay estuary, California. Quaternary Research 62: 134–145.

Maloney, A. B. 1936. Hudson's Bay Company in California. Oregon Historical Quarterly 37(1): 9–23.

Maloney, A. B., and J. Work. 1943. Fur brigade to the Bonaventura: John Work's California Expedition of 1832–33 for the Hudson's Bay Company. California Historical Society Quarterly 22: 323–348.

Mensing, S., and R. Byrne. 1998. Pre-mission invasion of *Erodium cicutarium* in California. Journal of Biogeography 25: 757–762.

Menzie, A., and A. Eastwood. 1924. Archibald Menzies' journal of the Vancouver Expedition. California Historical Society Quarterly 2: 265–340.

Miller, A. W., R. S. Miller, H. C. Cohen, and R. F. Schultze. 1975. Suisun Marsh Study, Solano County, California. U.S. Department of Agriculture Soil Conservation Service, Davis, California.

Milliken, R. 1991. An Ethnohistory of the Indian People of the San Francisco Bay Area from 1770 to 1810. Berkeley: University of California Press.

Milliken, R. 1995. A Time of Little Choice: The Disintegration of Tribal Culture in the San Francisco Bay Area 1769–1810. Menlo Park, California: Ballena Press.

Minnich, R. A. 2008. California's Fading Wildflowers: Lost Legacy and Biological Invasions. Berkeley: University of California Press.

Mitsch, W. J., and J. G. Gosselink. 2007. Wetlands, 4th ed. Hoboken, New Jersey: John Wiley.

Moffitt, J. 1938. Environmental factors affecting waterfowl in the Suisun area, California. Condor 40: 76–84.

Monda, M. J., and J. T. Ratti. 1988. Niche overlap and habitat use by sympatric duck broods in eastern Washington. Journal of Wildlife Management 52: 95–103.

Newberry, J. S. 1857. Reports on the Geology, Zoology, and Botany of Northern California and Oregon. War Department, Washington, D.C.

O'Rear, T. A., and P. B. Moyle. 2009. Trends in fish populations of Suisun Marsh January 2008—December 2008. Department of Wildlife, Fish, and Conservation Biology, University of California, Davis.

Ogden, A. 1933. Russian sea-otter and seal hunting on the California coast, 1803–1841. California Historical Society Quarterly 12: 217–239.

Patterson, K. B., and T. Runge. 2002. Smallpox and the Native American. American Journal of Medical Science 323: 216–222.

Preston, W. 1998. Serpent in the garden: environmental change in colonial California. Pages 260–298 in Contested Eden: California before the Gold Rush (R. A. Gutierrez and R. J. Orsi, eds.). Berkeley: University of California Press.

Rosenthal, J. S., G. G. White, and M. Q. Sutton. 2007. The Central Valley: a view from the catbird's seat. Pages 147–164 in California Prehistory: Colonization, Culture, and Complexity (T. L. Jones and K. A. Klar, eds.). Lanham, Maryland: Altamira Press.

San Francisco Estuary Institute. 2012. Bay Area EcoAtlas. http://www.sfei.org/ecoatlas/.

Schoolcraft, H. R., W. Clark, L. Cass, P. B. Porter, J. Monroe, D. Lowry, G. Gibbs, P. Prescott, D. D. Mitchell, S. M. Irvin, and others. 1857. Historical and Statistical Information Respecting the History, Condition and Prospects of the Indian Tribes of the United States. Philadelphia: Lippincott, Grambo.

Sloan, D. 2006. Geology of the San Francisco Bay Region. Berkeley: University of California Press.

Solano County Surveyor. 1920s. Maps of Solano County, California. On file with the Solano County Surveyor in Fairfield, stored on film.

Stone, A. 1996. Community is reborn by going back to its roots. USA Today, December 27.

Stoner, E. A. 1934. Summary of a record of duck shooting on the Suisun Marsh [California]. Condor 36: 105–107.

Stoner, E. A. 1937. A record of twenty-five years of wildfowl shooting on the Suisun Marsh, California. Condor 39: 242–248.

Sweeny, W. J., and G. H. Spencer. 1984. Plan of Protection for the Suisun Marsh including Environmental Impact Report. California Department of Water Resources, Sacramento.

Thompson, J. 1957. The Settlement Geography of the Sacramento–San Joaquin Delta, California. PhD dissertation, Stanford University, Stanford, California.

Treutlein, T. E. 1972. Fages as Explorer, 1769–1772. California Historical Quarterly 51: 338–356.

Unruh, J. R., and S. Hector. 2007. Subsurface characterization of the Potrero–Ryer Island thrust system, western Sacramento–San Joaquin Delta, Northern California. Pages 7–19 in Subsurface Characterization of the Potrero–Ryer Island Thrust System, Western Sacramento–San Joaquin Delta, Northern California. Bakersfield, California: American Association of Petroleum Geologists, Pacific Section.

U.S. Bureau of Reclamation, U.S. Fish and Wildlife Service, and California Department of Fish and Game. 2011. Suisun Marsh Habitat Management, Preservation, and Restoration Plan. Final environmental impact statement/environmental impact report. U.S. Bureau of Reclamation, Sacramento.

U.S. Geological Survey. 1896. Carquinez quadrangle, California. 1:62500. U.S. Geological Survey, Reston, Virginia.

U.S. Geological Survey. 1908. Antioch quadrangle, California. 1:62500. U.S. Geological Survey, Reston, Virginia.

U.S. Geological Survey. 1918a. Denverton quadrangle, California. 1:31680. U.S. Geological Survey, Reston, Virginia.

U.S. Geological Survey. 1918b. Fairfield south quadrangle, California. 1:24000. U.S. Geological Survey, Reston, Virginia.

U.S. Geological Survey. 1918c. Honker Bay quadrangle, California. 1:31680. U.S. Geological Survey, Reston, Virginia.

Von Langsdorff, G. H. 1814. Voyages and Travels in Various Parts of the World during the years of 1803, 1804, 1805, 1806, and 1807. London: Henry Colburn.

Wetland and Water Resources. 2011. Rush Ranch Existing Conditions Report. San Rafael, California: Wetlands and Water Resources.

Whipple, A. A., R. M. Grossinger, D. Rankin, B. Stanford, and R. A. Askevold. 2012. Sacramento–San Joaquin Delta historical ecological investigation: exploring pattern and process. San Francisco Estuary Institute Publication No. 627. Richmond, CA: San Francisco Estuary Institute–Aquatic Science Center.

GEOSPATIAL DATA SOURCES

Bache, A. D. 1872. Chart Number 626 Suisun Bay California. 1:40000. U.S. Coast Survey. National Oceanic and Atmospheric Association. Available: http://historicalcharts. noaa.gov/historicals/.

Boul, R., M. Vaghti, and T. Keeler-Wolf. 2009. Vegetation change detection of Suisun Marsh, Solano County, California: 1999, 2003, and 2006. California Native Plant Society.

Bowen, J. 2009. Old Stone Building Sparks New Historical Research and Discoveries. Solano Historical Society, Fairfield, California.

California Department of Water Resources Suisun Ecological Workgroup. 2007. LIDAR dataset. Available by request. Accessed: June 2012.

Camp, C. L., and G. C. Yount. 1923. The Chronicles of George C. Yount: California Pioneer of 1826. California Historical Society Quarterly 2: 3–66.

Eldredge, Z. S. 1909. The March of Portola and the Discovery of the Bay of San Francisco. San Francisco: California Promotion Committee.

Foxgrover, A., R. E. Smith, and B. E. Jaffe. 2012. Suisun Bay and Delta bathymetry. U.S. Geological Survey. Available: http://sfbay.wr.usgs.gov/sediment/delta/downloads.html. Accessed: June 2012.

Gesch, D., M. Oimoen, S. Greenlee, C. Nelson, M. Steuck, and D. Tyler. 2002. The National Elevation Dataset. Photogrammetric Engineering and Remote Sensing 68: 5–11.

Maloney, A. B., and J. Work. 1943. Fur brigade to the Bonaventura: John Work's California Expedition of 1832–33 for the Hudson's Bay Company. California Historical Society Quarterly 22: 323–348.

Metsker, C. F. 1953. Metsker's Map of Solano County, California. Available: California State Library (s68).

Milliken, R. 1995. A Time of Little Choice: The Disintegration of Tribal Culture in the San Francisco Bay Area 1769–1810. Menlo Park, California: Ballena Press.

San Francisco Estuary Institute. 2012. Bay Area EcoAtlas. Available: http://www.sfei.org/ecoatlas/. Accessed: March 20 2012.

Shumway, B. M. 1988. California Ranchos: Patented Private Land Grants Listed by County. San Bernardino, California: Borgo Press.

U.S. Bureau of Land Management. 1993. Public Land Survey. Available: http://www.geo-communicator.gov/GeoComm/lsis_home/home/. Accessed: 2003.

U.S. Geological Survey. 1896. Carquinez quadrangle, California. 1:62500. U.S. Geological Survey, Reston, Virginia.

U.S. Geological Survey. 1918a. Denverton quadrangle, California. 1:31680. U.S. Geological Survey, Reston, Virginia.

U.S. Geological Survey. 1918b. Fairfield South quadrangle, California. 1:24000. U.S. Geological Survey, Reston, Virginia.

U.S. Geological Survey. 1949 (1980). Fairfield South quadrangle, California. 1:24,000. U.S. Geological Survey, Reston, Virginia.

U.S. Geological Survey. 1953a (1980). Denverton quadrangle, California. 1:24,000. U.S. Geological Survey, Reston, Virginia.

U.S. Geological Survey. 1953b (1980). Honker Bay quadrangle, California. 1:24.000. U.S. Geological Survey, Reston, Virginia.

Whipple, A. A., R. M. Grossinger, D. Rankin, B. Stanford, and R. A. Askevold. 2012. Sacramento–San Joaquin Delta historical ecological investigation: exploring pattern and process. San Francisco Estuary Institute Publication No. 627. Richmond, CA: San Francisco Estuary Institute–Aquatic Science Center.

3

Physical Processes
and Geomorphic Features

Christopher Enright

Suisun Marsh is an uncommon place. Geologically, the Marsh is a very young landscape that occupies a widening of the Holocene river valley that drained the Central Valley between 10,000 and 3,000 years ago when sea level was about 2 m lower than today. Sea level rose rapidly up to around 6,000 years ago when tidal influence extended into the Delta and Suisun Marsh roughly to where it does today. Over the past 6,000 years sea level rise slowed to about 1–2 mm/yr, allowing sedimentation and erosion processes to form branching sinuous channels by the action of tidal currents and the interplay of sedimentation and vegetation at the edges (see chapter 2). By chance, the Marsh's roughly 470 km^2 occupy the geographic center of the northern reach of the San Francisco Estuary. This position alone accounts for much of its distinctive character, including its highly variable salinity. Today, the flows, exchanges, and morphology of Suisun Marsh are enormously modified from their historical condition. Yet the Marsh remains a key region in the life history of many estuarine species, even as further changes occur from climate change, sea-level rise, and seismic events. In this chapter, I discuss how the historical and modern interaction of tides, river flows, landscape morphology, and salinity control facilities make Suisun Marsh a distinctive part of the Estuary that is poised at the threshold of change.

TIDAL CURRENTS, RIVER FLOWS, AND SALINITY

Suisun Marsh experiences the most variable salinity patterns in the Estuary because it straddles the zone of most rapid west-to-east salinity change (figure 3.1; Moyle et al. 2010). The estuarine salinity gradient constantly changes under the dynamic tension between freshwater inflows from the Sacramento–San Joaquin

Delta watershed and ocean salt intrusion from the west. Because of the higher density of salt water, lighter fresh water tends to flow downstream over the top of heavier salt water that advances upstream at the bottom.

At the same time, the huge energy of two-a-day flooding and ebbing currents is dissipated by friction with the bottom, creating turbulent eddies and current shears that mix salt water upward and fresh water downward. In Suisun Bay, incoming tidal flows peak at 6,000 to 12,000 cubic meters/second (cms) only to reverse and flow out about 6.2 hours later. This interplay of tides, river inflows, and salt is most dynamic in Suisun Bay, where at one moment the water column can be quiescent with fresher water at the top and saltier water at the bottom, only to be roiled and mixed a few hours later. This rapid cycle of salinity stratification and mixing is a key physical influence on many estuarine species. Tidal mixing has more influence in shallow shoals and bays, whereas Suisun Bay's channels retain more river influence and maintain a degree of stratification even during extreme high and low tides, when water is moving most rapidly. This interplay of mixing and river flow is itself in constant motion as the strength of tidal flows move the salinity gradient 15 to 25 km each ebb and flood tide.

Averaged over the day, the west-to-east gradient of salinity varies smoothly from the Golden Gate to the limit of reversing tides near Walnut Grove on the Sacramento River and Vernalis on the San Joaquin River. Dilution of ocean salt begins slowly through Central and San Pablo bays and then increases rapidly through Suisun Bay before it gradually becomes fresh water at the periphery of the Delta. The location where the upstream end of the salinity gradient begins is often called the "low salinity zone" (LSZ), where peak abundances of several estuarine species such as delta smelt (*Hypomesus transpacificus*) and striped bass (*Morone saxatilis*) often occur (see figure 3.1).

Overlaid on rapidly changing tidal processes, the LSZ's position is also affected by slower processes such as changes in atmospheric pressure and lunar-cycle oscillations that deepen and then shallow the Estuary by up to one-third of a meter every two weeks. The largest influence on average location of the LSZ along Suisun Bay is seasonal variation in freshwater flows (Kimmerer 2004). During the dry season, freshwater flow is dwarfed by tidal currents in Suisun Bay, which can be 30 to 50 times greater than freshwater flow during peak flood or ebb tides. The salt balance in Suisun Bay is then dominated by tides, and the LSZ is often well east near the confluence of the Sacramento and San Joaquin rivers. Summer and fall marsh salinity is brackish, and western Suisun Marsh is often about one-half the salinity of sea water. During extended wet periods when the magnitudes of tidal and river currents are comparable, the LSZ approaches the western end of Suisun Bay, and eastern Suisun Marsh can freshen completely. The position of the LSZ is often indexed by "X2," defined as the distance from the Golden Gate upstream to the location of 2 ppt bottom salinity (X2 is also correlated with

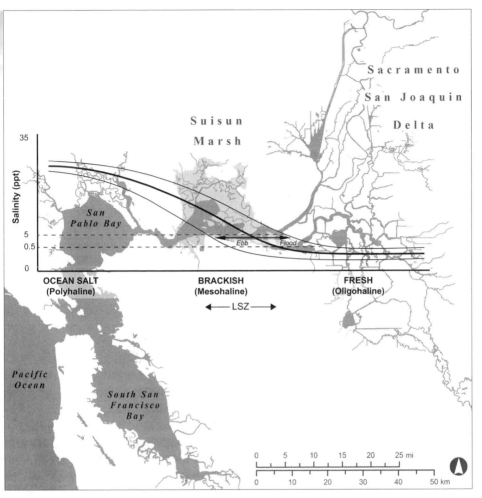

FIGURE 3.1. Representation of the tidal salinity gradient and position of the "low salinity zone" (LSZ) in the San Francisco Estuary.

abundances of phytoplankton and several aquatic invertebrate and fish species; Jassby et al. 1995).

The salinity of Suisun Marsh is also affected by freshwater inflow from the several small watersheds that enter the Marsh, primarily along its northern perimeter (see chapter 1). Inflows are very small, except for flash inflows during storms. Despite low inflow magnitudes, the tidal creeks that receive the flow are themselves small, so significant dilution can occur in upper tidal channels. Farther downstream, the freshening effect of creek inflows occurs only during ebb-tide pulses, especially on spring tides when the transition from the highest to the lowest tide is greatest. Salinity is also reduced in Boynton Slough by treated wastewater discharges, which average 0.75 cms (or 25 cubic feet/second [cfs]) from the Fairfield–Suisun Sewer District wastewater treatment plant (see map 6 in color insert). Overall, the effect of freshwater inputs from the northern creeks and the wastewater treatment plant creates a persistent salinity gradient from north to south along Suisun Slough.

THE INTERACTION OF TIDES
AND GEOMORPHOLOGY

Tides are the gravitational response of the world's oceans to the relative motions of sun, moon, and earth. The geomorphology of tidal marshes reflects the close interaction of tides (shaping the land) and the land (affecting tidal currents). Through sedimentation and vegetation processes, tidal dynamics are a key physical control on habitat structure and food webs for aquatic and semiaquatic organisms.

Tides

Like most coasts on the Earth, the San Francisco Estuary exhibits predominantly mixed semidiurnal tides, with two unequal high tides and two unequal low tides per lunar day (24.84 hours; Walters and Gartner 1985). The tidal range is nearly 1.2 m on neap tides and up to 2.0 m on spring tides. Every two weeks, the full or new moon aligns with the Sun and Earth, causing higher "spring" tides—especially in the summer and winter. Weaker "neap" tides occur about 7 days later, when the Sun, Moon, and Earth are in quadrature (i.e., near a 90° angle to each other).

High tides propagate from the Golden Gate to Suisun Marsh in about 3.5 hours, while low tides lag by about 4 hours. The difference is the result of waves traveling faster in deeper water, causing flood tides to propagate upstream more quickly than low tides recede. This differential effect of depth on tide propagation distorts the tidal pattern in Suisun compared with the regular harmonic pattern at the Golden Gate. There is also a consistent seasonal pattern in the timing of diurnal tide extremes. In recent decades (and for some decades to come), extreme high tides in Suisun Marsh occur near noon in the winter and near midnight in the summer. This sequence gradually inverts over a period of 167 years (Malamud-Roam 2000). All processes mediated by sunlight and flooding frequency are affected by this astronomical pattern. For example, bay shoals and mudflats are presently more exposed during the day in the summer, and natural tidal marshes tend to be flooded by high tides at night. The opposite pattern occurs in winter months. This pattern is important because it means, among other things, that summer water temperatures are periodically cooled near tidal marshes and that birds can forage mudflats during the day. A century or so hence, water temperature may increase near tidal marshes, and mudflat exposure will occur more at night (Enright et al. 2013).

Ebb–Flood Tidal Current Dominance

Estuarine tides are distorted by processes that compete to produce stronger flood or ebb currents (Friedrichs and Aubrey 1988). When the duration of ebb tide is shorter than that of flood tide, the ebb currents must be faster to compensate and the system is considered "ebb-dominant." When the opposite is true, the system is "flood-dominant" (Walton 2002). Sediment transport over the long term is sensitive to whether the fastest currents occur on ebb or flood tide, which strongly influences the morphology of marshes, channels, and bays (Friedrichs and Perry 2001). The deep channels of Suisun Bay generally maintain an ebb-dominant pattern because ocean tides at the Golden Gate sequence the low-low tide after the high-high tide about 80% of the time (Malamud-Roam 2000). Consequently, maximum tidal currents occur most often on ebb tides, with clear implications for transport of sediment and of planktonic plants and animals. By contrast, the morphology of natural tidal sloughs overcomes the astronomically driven tendency for ebb-dominance. Bank elevations of natural channels are dynamically maintained near mean high water level by balancing interactions among tide, vegetation, and sedimentation processes (Friedrichs and Perry 2001). This mechanism allows tidal marshes to adjust to rising sea level. About 15% of high tides flow over the top of natural channel banks in Suisun Marsh. At the moment of flood overtopping, the flow area suddenly expands, allowing flow and water velocity to increase substantially. Strong flood-dominance has been measured at First Mallard Branch (see map 6 in color insert), with flood currents exceeding 0.4 m/s on the highest summer and winter tides while the returning ebb-tide currents top out at about 0.25 m/s (Enright et al. 2013). Diked sloughs adjacent to managed wetlands no longer allow flood currents to flow over banks, leaving those sloughs always ebb-dominant. *This loss of tidal connectivity over much of Suisun Marsh has removed the primary mechanism by which the landscape can respond to sea-level rise.*

Tidal Flow, Excursion, and Prism

The proximity of expansive bays, large tidal channels, and small tidal creeks creates extremely variable tidal currents connecting diverse regions of Suisun Marsh. The rates at which water is exchanged within Suisun Marsh and between the Marsh and Suisun Bay largely determine how sediment, food-web production, and nekton are shared among the regions. Suisun Bay itself is highly variable, from the western end, where tidal flows average ±17,000 cms, to the eastern end near Chipps Island, where friction imparted by Bay and Marsh reduce the

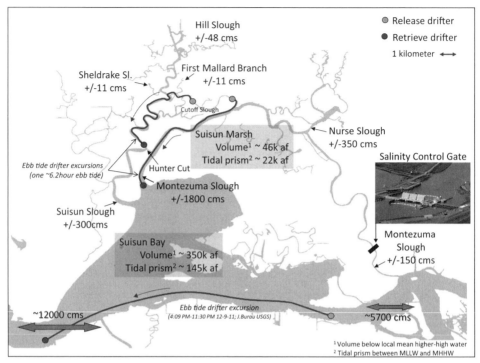

FIGURE 3.2. Tidal flows, exchanges, and prisms in Suisun Marsh. Heavy lines show the paths of objects allowed to drift with the tide; light gray dots show the start, and dark gray dots the end, of each pathway over one 6-hour ebb tide.

range of tidal flows to about ±8,500 cms (see figure 3.2). The approximately 6.2-hour tidal excursion of a passive "drifter" released on slack after flood tide at Chipps Island traveled 22 km to Martinez in Carquinez Strait (J. Burau [U.S. Geological Survey], personal communication, 2012). About 40% of the water in Suisun Bay enters or leaves it on the strongest tide of each day. Contemporary Suisun Marsh connects to Suisun Bay only at the ends of Montezuma Slough and the mouth of Suisun Slough. Most of the tidal energy and water exchange occurs at the western entrance, where Montezuma Slough tidal flow is ±1,850 cms and Suisun Slough is more than ±300 cms. Hunter Cut was constructed in the 1930s between Montezuma and Suisun sloughs. The difference in tide arrival time at either end of Hunter Cut drives strong tidal flows averaging ±1,400 cms. Current velocity is briefly very high on each tide (more than 1 m/s), maintaining Hunter Cut as the deepest slough in Suisun Marsh (about 11 m). The eastern Montezuma Slough entrance near Collinsville is much less energetic, conveying ±150 cms each tide unless the Suisun Marsh Salinity Control Gate (SMSCG) is operating (see below). Tidal-creek watersheds around the perimeter of Suisun Marsh have tidal flows that scale to the size of the creek system. Figure 3.2 shows maximum flows at the mouth of several tidal-creek watersheds around the Marsh. The largest is Nurse Slough, which conveys nearly 350 cms each tide in a creek network with about 3.2 million m^3 of volume below mean sea level. Tidal excursions in Suisun Marsh depend on where observations are made. Where water movement is not constrained by dead-end channels, excursions can encompass large areas of the Marsh. Passive drifters released on ebb-tide excursion from either side of the tidal convergence on Cutoff Slough traveled 8 to 10 km along Suisun and Montezuma sloughs (figure 3.2). The tidal-cycle volume exchange—or "tidal prism"—between Suisun Marsh and Suisun Bay can be estimated from continuous flow measurement at the three Suisun Bay connections. On average, about 27 million m^3 of water is exchanged with Suisun Bay. This amounts to nearly 40% of the volume of Suisun Marsh below the mean high water level.

METEOROLOGICAL DRIVERS

Along with river inputs and tides, meteorological factors such as wind, evaporation, and changes in barometric pressure also drive flows. These forces are generally covarying, but I present them one at a time and discuss their interactions.

Wind

The prevailing winds in Suisun Marsh are from the west and southwest. When barometric pressure is low, wind is generally southwesterly. The wind is out of the north approximately 10% of the time, most often in winter during fair weather. Because most Marsh channels meander, reaches that are oriented along the wind direction experience waves and high turbidity during wind events. In reaches where channel orientation is more perpendicular to the wind, turbidity is lower and the water surface is far more quiescent. At the regional scale, over 40% of the subtidal variation in Suisun Bay sea level is caused by regional winds acting on coastal ocean sea level (Ryan and Noble 2007). The prevailing westerly winds from the coastal ocean at the Golden Gate can drag water upstream and increase tide heights in Suisun Bay as a result of persistent wind shear at the water surface.

Evaporation

Suisun Marsh contains a mosaic of emergent aquatic vegetation that is highly efficient at transpiring water. Tule, tuberous bulrush, cattail, and rush species can transpire about ⅔ m to 3 m of water per unit area in a year, depending on temperature, wind, and humidity. During late spring and summer, net flows in terminal sloughs are slightly negative (i.e., flood-directed), owing to evaporation losses from surface water and vegetation along the slough or in adjacent marsh plains. During the warm months, sloughs within tidal marshes become water "sinks" as evaporation at the upper ends causes a net upstream flow. For example, an undiked slough, First Mallard Branch, accumulates water rapidly during spring tides when channel water overflows the bank and occupies the marsh plain where it is easily evaporated. Evaporative processes carry away heat by several mechanisms, making terminal slough systems important temperature modulators during summer months (Enright et al. 2013).

Barometric Pressure

Barometric pressure often covaries with wind direction and evaporation. During storm events, barometric pressure can drop rapidly, which increases sea level in the coastal ocean outside the Golden Gate. The potential for levee damage and flooding of subsided areas in Suisun Marsh increases markedly when storms coincide with astronomical high tides. There are often two peaks in average water levels in the wake of storms. As the storm passes over, the down spike in barometric pressure can raise average sea level by 0.2–0.3 m. In the days after the storm passes and barometric pressure increases, another somewhat lower peak in average Suisun Bay level occurs as the pulse of Delta outflow moves through (Walters and Gartner 1985).

PHYSICAL GEOMETRY OF SUISUN MARSH

Historical Morphology

The geomorphology of Suisun Marsh is highly modified from its historical condition. An 1883 Suisun Bay navigation chart indicates some of the changes (see map 7 in color insert). In contrast to today, the southern named islands were actual islands with narrow but relatively deep channels between. We can suppose that tidal currents in these channels would often be swift, especially when tidal heights from one end to the other were unequal. This was likely the case for "Roaring River," which connected Grizzly Bay to eastern Montezuma Slough and which perhaps earned its name because high tides reach each end of the slough about 40 minutes apart. This difference in water levels over a relatively short distance can drive significant flows and would also tend to mix the Suisun Bay salinity gradient. The remarkable detail of the historical chart also reveals the extensive connectivity between Suisun Bay and southern island tidal marshes via small tidal creek systems. Observations of the few remaining natural tidal creek systems in Suisun, like those at Rush Ranch, give a window on what was historically a common geomorphic feature. The bay-connecting creeks are the downstream ends of complex tidal-creek networks that were dendritic in shape and had several orders of branching before they terminated on the marsh plain. Within the creek systems, water and material exchanges were extremely variable.

The smallest branches received water only on high tides and were otherwise dry, even though groundwater was very near the surface. Connectivity to Suisun Bay was thus episodic. For example, small fish seeking small-creek-branch refuges and food could only find access for hours at a time. Larger downstream branches had longer periods of water retention, although they might still dewater at low tides. The largest slough branches that connected directly to Bay or island sloughs exchanged almost all of their water each tide. All creek-system branches had steep banks that were generally undercut, resulting in intermittent bank slumps. Creek-branch cross sections narrowed with distance, keeping water velocity nearly constant from one end of a branch to the other (Lawrence et al. 2004). Connections to smaller creeks stepped up in elevation at the branch entrance. Only the highest tides flowed over banks, where water velocity decreased rapidly and suspended sediments settled out quickly. This process gave soils at natural channel edges more mineral content and good drainage that supported deep-rooting vegetation such as tules and cattails. Inorganic sediment deposition decreased with distance from bank edges, so the spatial distribution of tidal creeks largely controlled the distribution of soil characteristics and, in turn, the distribution of plant and animal communities.

Sea-level rise was accommodated naturally by these processes. Soils in the Marsh interior comprise deposited riverborne mineral sediments and organic matter from above- and belowground vegetation. As sea level rose, a positive feedback was generally maintained between mineral sediment deposits that support vegetation, and vegetation structure that supported further sediment trapping.

Physical Geometry of Suisun Marsh Today

Natural tidal marshes are rare in Suisun Marsh today. Even the best examples of natural tidal marshes at Rush Ranch have been modified significantly by historical cattle grazing, by burning for vegetation management, and by artificial ditching for mosquito control. Nevertheless, Suisun remains a dynamic landscape of contrasts and potential for significant change. Categories of physical geometry include aquatic features such as bays, channels, sloughs, and upland creeks. Terrestrial features include tidal marshes, tidal fringes, diked public and private lands, and infrastructure. The slow influence of sea-level rise and the episodic influence of more extreme tides and seismic events will likely change the fundamental character of Suisun Marsh in coming decades. It is a landscape poised on the edge of major changes, both planned and unplanned.

Bays, Channels, Sloughs, Creeks

Contemporary Suisun Marsh is very much changed from its historical condition, although it still contains a complex of channels and sloughs along with extensive water-control structures (see map 6). Of the 485 km^2 (120,000 acres) encompassed by Suisun Marsh, approximately 105 km² is bays, channels, and sloughs. Montezuma Slough spans the Marsh for 34 km (21 mi), connecting with the lower Sacramento River on the east and shallow Grizzly Bay on the west. The longest terminal slough in the Marsh is Suisun Slough, which meanders 23 km (14 mi) from Suisun City in the north toward the south, where it connects to western Grizzly Bay. Several smaller slough systems radiate from Montezuma and Suisun sloughs, including the Nurse Slough complex in the northeast Marsh, the Hill Slough complex in the north-central Marsh, Peytonia and Boynton sloughs in the northwest Marsh, and the Suisun and Green Valley Creek watersheds in the western Marsh (see map 6). Most Suisun Marsh channels and sloughs are bordered by levees protecting managed wetlands. Dendritic tidal creeks, once the dominant feature of marsh plain topography, are now mostly cut off by the slough levees. Their remnants can be seen in vegetation patterns on managed wetlands where soils contain more inorganic sediment, a relic of the past geomorphic response to sea-level rise. Many larger slough complexes have been leveed across their entrance and connected only through culverts. Examples include the mouth of Noyce Slough on Simmons Island, Frost Slough on Grizzly Island, and Volanti Slough on Joice Island. These sloughs are now maintained as nontidal lakes that serve to distribute and control water levels in managed wetlands. A rough estimate is that the lost land-to-water interface once encompassed by small sloughs and tidal creeks reaches into the hundreds of miles in Suisun Marsh.

Fringe Marshes

Fringe marshes are long, narrow bands of emergent vegetation along wetland banks. In Suisun, they are artifacts of constructed levees that were set back from the channel edge for stability and protection from wind and waves. They vary in size and species composition, with little geomorphic complexity and low area-to-edge ratio. The best examples of fringe marshes are in the Nurse Slough complex, where levees historically were constructed with significant setback. Some fringe marshes have expanded laterally over time wherever sediment has been deposited after changes in nearby physical geometry. An example is lower Suisun Slough, which lost tidal current energy after Hunter Cut was constructed in the 1930s. It has filled with sediment, and tidal fringes have expanded laterally. Because they are dominated by emergent vegetation, fringe marshes provide surface area for epiphytic algae, structural refuge for fishes, and, in some places, significant water storage and residence time that influence flow dynamics in the slough.

Managed Wetlands

About 210 km² (52,000 acres) of Suisun Marsh comprises wetlands managed primarily for waterfowl habitat and hunting; 144 km² (35,700 acres) of this area belongs to private duck clubs, while 62 km² (15,300 acres) is owned by

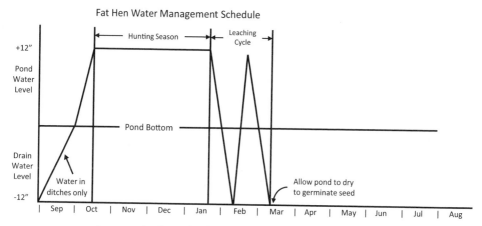

FIGURE 3.3. An example of a wetland water-management strategy.

the California Department of Fish and Wildlife. Over 350 km (210 mi) of constructed levees separates managed marshes from tidal sloughs. Water is managed to control salt accumulation in soil and promote growth of waterfowl food plants within some seasonal regulatory bounds. Wetland managers construct ditches to topographically low areas and use disking and contouring to minimize evaporation and salt accumulation in the soil. The tides and over 300 gated culverts are used to control seasonal flooding and drainage of land. Duration and depth of flooding are managed to attract waterfowl. Flooding typically occurs in September and October, when high tides and gravity fill distribution ditches and ponds. Several plant-growth strategies are used to encourage establishment of different species. Figure 3.3 shows an example strategy for growing fathen (*Atriplex prostrata*) (Rollins 1981). During the hunting season, culvert flashboards control pond water level, typically to a depth of 12 inches to attract dabbling ducks such as mallard and pintail. Water is circulated during hunting season to remove excess organic matter. The hunting season ends in late January, and ponds are then drained. Depending on the management strategy, one or more "leach cycles" fill and drain the pond quickly to remove salt from the soil surface. Pumps are often required to drain water because subsided land does not allow gravity drainage at low tide.

SALINITY CONTROL IN SUISUN MARSH TODAY

Salinity is a key habitat characteristic that affects both physical and ecological processes in the Suisun Marsh–Bay system. As discussed above, salinity is highly variable in Suisun owing to the position of the Marsh along the estuarine salinity gradient. The location of the salinity gradient can shift 15 to 25 km (9–16 mi) within 1 day on the tidal ebb and flow (see figure 3.1). The long-term average position of the salinity gradient depends primarily on Delta outflow. Because Delta outflow is reduced by state and federal water-project exports from the south Delta, salinity control has been the primary water management objective in Suisun Marsh since the 1980s, when salinity control facilities were constructed.

Salinity Control Structures

State and federal water projects built several salinity control structures in the Marsh between 1980 and 1990, including the SMSCG, Roaring River Distribution System, Morrow Island Distribution System, and Goodyear Slough Outfall (map 6 in color insert). Constructed in 1988, the SMSCG is the largest and most effective salinity control structure within the Marsh. The SMSCG is located on Montezuma Slough, about 3 km downstream from the confluence of the Sacramento and San Joaquin rivers near Collinsville. The purpose of the SMSCG is to decrease the salinity of the water in Montezuma Slough to meet salinity standards set by the State Water Resources Control Board at several monitoring stations in Suisun Marsh. The facility, spanning the 140 m (465 ft) width of Montezuma Slough, consists of a boat lock, three operable radial gates, and removable flashboards. The gates control salinity by restricting the flood-tide flow of higher-salinity water from Grizzly Bay into Montezuma Slough (radial gates closed) and retaining lower-salinity Sacramento River water from the previous ebb tide (radial gates open). Operation of the gates in this fashion lowers salinity in Suisun Marsh channels and results in a net movement of water from east to west.

When Delta outflow is low to moderate and the SMSCG is not operating, tidal flow past the gate is approximately ±150 cms (5,300 cfs), while the net flow—the average over the tides—is near zero. When the gate is operated, flood-tide flows are arrested while ebb-tide peak flows remain in the range of 150 cms. The net flow in Montezuma Slough becomes approximately 80 cms (2,800 cfs) (see figure 3.4).

The Army Corps of Engineers permit for operating the SMSCG allows operation between October and May only when needed to meet Suisun Marsh salinity standards. In many years, the gate has been operated as early as October 1, although in some wet years (e.g., 2006) the gate was not operated at all. When the channel water's salinity decreases sufficiently below the salinity standards, or at the end of the control season, the flashboards are removed and the radial gates are raised to allow unrestricted flow through Montezuma Slough. The approximately 80 cms (2,800 cfs) net flow induced by SMSCG operation is highly effective in reducing salinity downstream in Montezuma Slough. Salinity is reduced by roughly one-half at Beldons Landing on Montezuma Slough (see figure 3.5), and by lesser amounts farther west along Montezuma Slough. At the same time, the salinity field in Suisun Bay moves upstream approximately 3 km (2 mi) as net

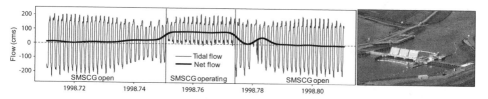

FIGURE 3.4. Montezuma Slough flow near the Suisun Marsh Salinity Control Gate (SMSCG). The gate was operated (open on ebb, closed on flood tide) during October 1–9, 1998. Black line shows the tidally averaged net flow. Net flow without gate operation is near zero, whereas net flow with gate operation is approximately 80 cms (~2,800 cfs).

Delta outflow (measured nominally at Chipps Island) is reduced by gate operation (figure 3.5). Net outflow through Carquinez Strait is not affected.

PHYSICAL PROCESS CHALLENGES AND OPPORTUNITIES FOR RESTORATION

The contemporary Suisun Marsh landscape presents many challenges and opportunities for restoration of tidal marsh functions. Several current initiatives to protect and restore Suisun Marsh share common goals to conserve native species and communities by restoring landscape processes and features reminiscent of the historical landscape. The fundamental premise is that native species are adapted to the historical climate, hydrology, community ecology, and landscape patterns. Challenges to these goals are many because climate is changing, hydrology is modified for water supply, and the community ecology and landscape pattern are altered from natural conditions by historical land use. According to Teal et al. (2009), the wetland restoration challenge has dual imperatives. First, the approach to ecosystem restoration should rely on the landscape's ability to self-repair. That is to say, work with natural processes—primarily tidal energy, sedimentation, and vegetation processes—and allow them to restore landscape patterns and ecosystem functions. Careful choices that minimally modify levees, control vegetation, or shape landforms will encourage natural processes that, over time, will recover structures and functions beneficial to native species. This approach minimizes economic costs and affords the best chance of successful and resilient outcomes. Second, Teal et al. (2009) encourage bold actions to recover natural processes and ecosystem functions before land subsidence and sea-level rise remove options and limit chances for recovery of native species functions. Self-repair of natural landscape patterns and functions must overcome significant challenges, including sea-level rise, land subsidence, low sediment supply, and decreased tidal energy.

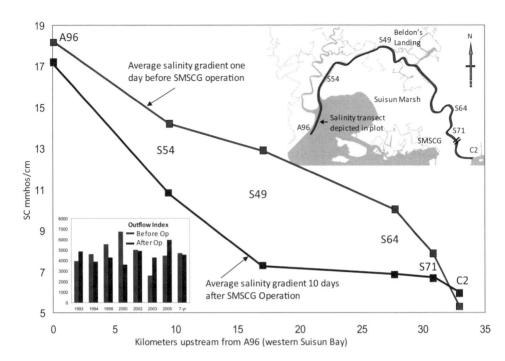

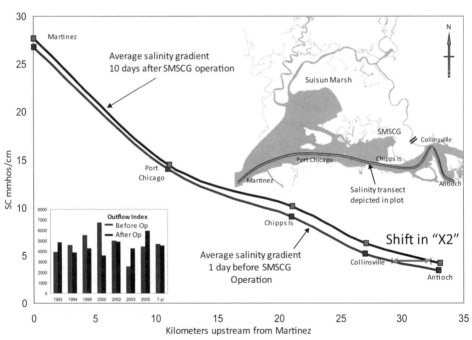

Challenge 1: Sea-level Rise

Sea level is currently rising more than 3 mm each year as a result of melting continental glaciers and thermal expansion of the world's oceans. This rate is increasing with time (Cayan et al. 2008, 2009; figure 3.6). Water and ecosystem restoration planning documents often use an estimate of 1.4 m (4 ft) of rise by 2100 (e.g., Bay Conservation and Development Commission 2008, 2010). Beyond average sea-level rise, planning for levee protection must consider extremes of water level because these events define the ultimate utility of levees. Levees in Suisun Marsh are vulnerable to the co-occurrence of high tides and storm conditions when low barometric pressure and winds can amplify high tides. Climate change is expected to increase the frequency of extreme tides in the future because conditions may bring more intense storms to California (Cayan et al., 2008).

Challenge 2: Land Subsidence

More than half the land in Suisun Marsh is currently below intertidal elevation (see maps 8 and 9 in color insert). With all else constant, 1.4 m of sea-level rise by 2100 would likely preclude options for natural elevation recovery. Much of Suisun Marsh is therefore near a geomorphic threshold between alternative futures. Depending on conditions and restoration approach, much of Suisun land elevation today is recoverable by natural processes over time, through positive feedbacks between sedimentation and vegetation processes. However, land uses that cause additional subsidence will change the dynamics gradually so that erosion processes—primarily due to wind and currents—will begin to predominate over accumulation processes. Restoration designs will need to consider options for allowing tidal exchange while encouraging sedimentation and vegetation feedbacks to recover elevation over time.

Challenge 3: Sediment Supply

Suspended sediment discharge to the San Francisco Estuary is decreasing, primarily as a result of upstream reservoir capture (Wright and Schoellhamer 2004). While highly variable, the average watershed sediment discharge to the Estuary is about 7 million m^3 (Krone 1979). By comparison, the subsided "accommodation space" (Mount and Twiss 2005) in Suisun is more than 300 million m^3 of soil volume. The potential for elevation recovery therefore depends on the interaction

FIGURE 3.5. *(opposite)* Average of 7 years of salinity response to the Suisun Marsh Salinity Control Gate (SMSCG) operation in Montezuma Slough and Suisun Bay. Gray line is salinity profile one day before gate operation; black line is salinity 10 days after gate operation. The upper chart depicts a transect from west to east along Montezuma Slough, the lower chart along Suisun Bay.

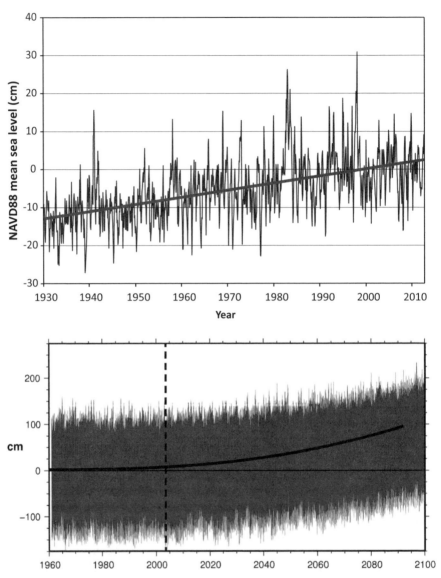

FIGURE 3.6. Historical sea-level rise in San Francisco Bay (top; National Oceanic and Atmospheric Administration 2013) and projected sea-level rise through 2100 (bottom; Cayan el al. 2009).

of several physical processes related to sediment supply, including (1) the initial land elevation (i.e., which side of the geomorphic threshold is the land on?); (2) physical forcing in the region, primarily by currents and wind; (3) proximity to sediment supply; (4) restoration design that leverages ebb–flood asymmetry to accumulate sediment; and (5) the potential for positive feedback between vegetation and sedimentation on the subsided restoration site.

Challenge 4: Tidal Energy

Much of the ecosystem functioning of restored tidal marshes comes from the extent of land area that is intertidal in elevation. At the same time, marsh areas dissipate tidal energy through friction in the shallow regions and through the interaction of flows with the vegetation. It is thus a particular challenge to expand intertidal wetlands when the tidal range is reduced in the process. The interconnected nature of hydrodynamic systems means that when structures are changed at one location, dynamics are altered elsewhere in the system. When pursuing large-scale restoration in Suisun Marsh and the Delta, which may involve removing or altering levees, it is critical to understand how actions at one location may affect intertidal extents throughout the landscape. Although climate change may somewhat increase the energy of tides sloshing back and forth in the system, total tidal energy in the San Francisco Estuary is basically a zero sum game. The more that tidal energy is dissipated, the less energy there is available for erosion and deposition.

ASSESSING THE FUTURE OF PHYSICAL PROCESSES IN SUISUN MARSH

The morphology and ecology of Suisun Marsh reflect the influence of hydrodynamic processes on currents, salt and sediment transport, and vegetation. Challenges posed by tidal energy, sea-level rise, sediment supply, and land subsidence are daunting. Of the four challenges described above, sea-level rise and sediment supply are global or legacy stressors that will not be influenced by restoration initiatives but must nevertheless be anticipated. By contrast, tidal energy and subsidence are challenges that are squarely influenced by current land use and wetland ecosystem-restoration choices. Land subsidence is tied to the 130-year history of Suisun Marsh as a waterfowl hunting area (see chapter 5). The traditional management practices keep wetland soils dry during the heat of summer and assure that the organic components oxidize over time. This practice continues in full view of the future negative consequences for sustaining traditional management practices and waterfowl conservation. Planning for redistribution of tidal energy will be on the frontier of restoration science. While tidal marsh restoration of subsided lands will dissipate tidal range and limit intertidal extents in the short term, the soil-building processes set in motion can reduce tidal energy dissipation over the long term. Ecosystem functions can be enhanced with careful choices about locations and extents of tidal marsh restoration along the northern reach, how tidal connections are made and staged with time. Hydrodynamics and transport models will be increasingly important for planning and design of restoration projects along the way.

REFERENCES

Bay Conservation and Development Commission. 2008. Living with a rising bay: vulnerability and adaptation in San Francisco Bay and on the shoreline. Bay Conservation and Development Commission, San Francisco.

Bay Conservation and Development Commission. 2010. Highlights of the Bay Delta Conservation Plan. San Francisco: Bay Conservation and Development Commission. http://baydeltaconservationplan.com.

Cayan, D. R., P. D. Bromirski, K. Hayhoe, M. Tyree, M. D. Dettinger, and R. E. Flick. 2008. Climate change projections of sea level extremes along the California Coast. Climatic Change 87 (Supplement 1): S57–S73.

Cayan, D., M. Tyree, M. Dettinger, H. Hidalgo, T. Das, E. Maurer, P. Bromirski, N. Graham, and R. Flick. 2009. Climate change scenarios and sea level rise estimates for the California 2008 Climate Change Scenarios Assessment. Draft Paper, California Energy Commission CEC-500-2009-014-D.

Enright, C., S. Culberson, and J. R. Burau. 2013. Broad timescale forcing and geomorphic mediation of tidal marsh water flow and temperature dynamics. Estuaries and Coasts. http://dx.doi.org/10.1007/s12237-013-9639-7.

Friedrichs, C. T., and D. G. Aubrey. 1988. Tidal distortion in shallow well-mixed estuaries: a synthesis. Estuarine, Coastal and Shelf Science 27: 521–545.

Friedrichs, C. T., and J. E. Perry 2001. Tidal salt marsh morphodynamics. Journal of Coastal Research (Special Issue) 27: 7–37.

Jassby, A. D., W. J. Kimmerer, S. G. Monismith, C. Armor, J. E. Cloern, T. M. Powell, J. R. Schubel, and T. J. Vendlinski. 1995. Isohaline position as a habitat indicator for estuarine populations. Ecological Applications 5:272–289.

Kimmerer, W. J. 2004. Open water processes of the San Francisco Estuary: from physical forcing to biological responses. San Francisco Estuary and Watershed Science [online serial]. Vol. 2, Issue 1 (February 2004), Article 1. http://repositories.cdlib.org/jmie/sfews/vol2/iss1/art1

Knowles, N. 2010. Potential inundation due to rising sea levels in the San Francisco Bay region. San Francisco Estuary and Watershed Science 8(1).

Krone, R. B. 1979. Sedimentation in the San Francisco Bay system. Pages 63–67 *in* San Francisco Bay: The Urbanized Estuary (T. J. Conomos, ed.). San Francisco: AAAS Pacific Division.

Lawrence, D. S. L., J. R. L. Allen, and G. M. Havelock. 2004. Saltmarsh morphodynamics: an investigation of tidal flows and marsh channel equilibrium. Journal of Coastal Research 20: 301–316.

Malamud-Roam, K. P. 2000. Tidal regimes and tide marsh hydroperiod in the San Francisco Estuary: theory and implications for ecological restoration. PhD dissertation, University of California, Berkeley.

Mount, J., and R. Twiss. 2005. Subsidence, sea level rise, and seismicity in the Sacramento–San Joaquin Delta. San Francisco Estuary and Watershed Science 3(1), Article 5.

Moyle, P. B., W. A. Bennett, W. E. Fleenor, and J. R. Lund. 2010. Habitat variability and complexity in the upper San Francisco Estuary. San Francisco Estuary and Watershed Science 8(3): 1–24. http://escholarship.org/uc/item/0kf0d32x.

National Oceanic and Atmospheric Administration. 2013. Tides and currents. http://tidesandcurrents.noaa.gov/. Accessed: March 2013.

Rollins, G. L. 1981. A Guide to Waterfowl Habitat Management in Suisun Marsh. California Department of Fish and Game, Sacramento.

Ryan, H. G., and N. A. Noble 2007. Sea level fluctuations in central California at subtidal to decadal and longer time scales with implications for San Francisco Bay, California. Estuarine, Coastal, and Shelf Science 73: 538–550.

Teal, J. M., N. G. Aumen, J. E. Cloern, K. Rodriguez, and J. A. Wiens. 2009. Ecosystem restoration workshop panel report. CALFED Science Program, Sacramento.

Walters, R. A., and J. W. Gartner. 1985. Subtidal sea level and current variations in the northern reach of San Francisco Bay. Estuarine, Coastal and Shelf Science 21: 17–32.

Walton, T. L. 2002. Tidal velocity asymmetry at inlets. ERDC/CHL CHETN IV-47. U.S. Army Engineer Research and Development Center, Vicksburg, Mississippi. http://chl.erdc.usace.army.mil.

Wright, S. A., and D. H. Schoellhamer. 2004. Trends in the sediment yield of the Sacramento River, California, 1957–2001. San Francisco Estuary and Watershed Science 2 (2). http://repositories.cdlib.org/jmie/sfews/vol2/iss2/art2.

GEOSPATIAL DATA SOURCES

Bache, A. D. 1872. Chart Number 626 Suisun Bay California. 1:40,000. United States Coast Survey (USCS). Courtesy of the National Oceanic and Atmospheric Administration. Available: http://historicalcharts.noaa.gov/historicals/.

CalAtlas. 2012. California Geospatial Clearinghouse. State of California. Available: http://atlas.ca.gov/. Accessed: 2012.

California Department of Water Resources. 2013. California Data Exchange Center. Available: http://cdec.water.ca.gov/. Accessed: March 2013.

California Department of Water Resources Suisun Ecological Workgroup. 2007. LIDAR dataset. Available by request. Accessed: June 2012.

Contra Costa County. 2013. Contra Costa County Mapping Information Center. Available: http://www.ccmap.us/. Accessed: January 2013.

Foxgrover, A., R. E. Smith, and B. E. Jaffe. 2012. Suisun Bay and Delta Bathymetry. U.S. Geological Survey. Available: http://sfbay.wr.usgs.gov/sediment/delta/downloads.html. Accessed: June 2012.

Gesch, D., M. Oimoen, S. Greenlee, C. Nelson, M. Steuck, and D. Tyler. 2002. The National Elevation Dataset. Photogrammetric Engineering and Remote Sensing 68 (1): 5–11.

San Francisco Estuary Institute. 2012. Bay Area EcoAtlas. Available: http://www.sfei.org/ ecoatlas/. Accessed: March 20 2012.

Solano County. 2013. Geographic Information Systems Homepage. Solano County Department of Information Technology. Available: http://www.co.solano.ca.us/depts/ doit/gis/. Accessed: November 2012.

4

Shifting Mosaics:
Vegetation of Suisun Marsh

Brenda J. Grewell, Peter R. Baye, and Peggy L. Fiedler

To those who have experienced Suisun Marsh, mere mention of this vast wetland conjures images of saltgrass–pickleweed plains and robust sedges that sway to the Suisun winds and tides along a complex labyrinth of tidal sloughs and shorelines. Upon closer look, the richness and diversity of the tidal flora and the marsh-dependent life it supports are revealed. Shifting mosaics of vegetation are a defining feature of Suisun wetlands. Seasonal changes of the plant communities are highly anticipated, but less obvious are changes to vegetation structure and composition that have varied over millennia and will continue to vary with time. Historically, variability in salinity and hydrologic processes was a key driver of biological diversity (Moyle et al. 2010), but these are less dynamic spatially and more dynamic temporally than they once were, owing to the development of extensive water management infrastructure (chapter 3; Enright and Culberson 2009). The nature of the estuarine vegetation's response to changing environmental conditions in the 21st century is a critical question for conservation management.

Estuarine wetlands vary considerably from the Golden Gate to the Sacramento–San Joaquin Delta, and vegetation within this west-to-east transect in Suisun Marsh is distinctive (Mason 1972; Atwater et al. 1979; Baye et al. 2000; Grewell et al. 2007). Tidally delivered ocean water blends with freshwater runoff so that plant species adapted to wide fluctuations in salinity and hydrology flourish in the Marsh. The largely azonal, patchy mosaic of tidal wetland vegetation is distinctive from vegetation in marshes downstream, where low-diversity salt marshes that have classic, relatively sharp vegetation zones in discrete tidal-elevation ranges prevail. Upstream, tidal influence occurs throughout the Delta, but freshwater conditions are maintained and emergent marsh-willow thicket-riverine forests diverge from the vegetation of the Marsh (Mason 1972; Hickson and Keeler-Wolf 2007). Freshwater outflow from the Delta historically has contributed to brackish conditions in wetlands throughout the Marsh. This is quite different from the salt marshes of San Francisco and San Pablo bays, where brackish marshes are confined to limited reaches of tidal sloughs where local freshwater discharges dilute higher-salinity bay water. See table 4.1 for a list of endemic and threatened species of Suisun Marsh.

Modification of the timing and magnitude of seasonal freshwater outflow and the salinity and flow regimes within estuaries can alter the character of hydrogeomorphic habitat and affect the distribution and abundance of associated biota (Simenstad et al. 2000a). Approximately 84% of the estuarine wetlands in Suisun Marsh are currently diked (San Francisco Estuary Institute 1998), with seasonal flood and drain cycles managed for waterfowl habitat. Long-term tidal restriction and partial drainage—and the subsidence, acid sulfate soil, and salinization that result—largely explain the composition and structure of emergent vegetation behind the dikes.

The future of Suisun Marsh poses many challenges. Temperature increases, more extended droughts, diminished snow pack, sea-level rise, and increased salinity intrusion are predicted (chapter 3; Knowles and Cayan 2002, 2004; Cayan et al. 2008a, 2008b; Cloern et al. 2011) and suggest that vegetation in the Marsh will experience longer hydroperiods and pulses of increased salinity. Plant species that thrive in conditions of sustained fresh water or regular drainage will likely decrease in abundance and may be limited to the upper edges of the San Francisco Estuary.

In this chapter, we describe contemporary plant communities by the hydrogeomorphic settings that link flora to physical processes in Suisun Marsh. We then describe the major environmental factors that drive the distribution and abundance of the vegetation and review paleobotanical evidence of long-term vegetation changes from the late Holocene to historical times. Looking forward, we project how plant communities and ecosystem-level properties may respond to predicted changes in physical processes.

PRINCIPAL WETLAND VEGETATION: LANDFORM ASSOCIATIONS

In Suisun Marsh, estuarine wetlands dominated by vascular plants are distributed across heterogeneous environmental conditions. Rather than using vegetation associations based on dominant species that occur in relatively small patches (e.g., Keeler-Wolf et al. 2000), we consider the vegetation of Suisun Marsh in very broad, practical units familiar from wetland classification systems; these systems

TABLE 4.1 Special-status plant taxa within estuarine wetlands, wetland–terrestrial ecotones, and terrestrial habitat of the Suisun Marsh region.

Family	Taxa	Common names	Listing status USA	State	CNPS
Apiaceae (Carrot)	*Cicuta maculata* var. *bolanderi* (S. Watson) G.A. Mulligan	Bolander's water hemlock			2
Asteraceae (Sunflower)	*Cirsium hydrophilum* (Greene) Jeps. var. *hydrophilum*	Suisun thistle	FE		
	Lasthenia conjugens Greene	Contra Costa goldfields	FE		
	Isocoma arguta Greene	Carquinez goldenbush		SC	1B
	Symphyotrichum lentum (Greene) G.L. Nesom	Suisun Marsh aster		SC	1B
Chenopodiaceae (Goosefoot)	*Atriplex cordulata* Jepson	Heartscale			1B
	Atriplex parishii S. Watson var. *depressa* Jepson	Brittlescale			1B
	Atriplex joaquinana (A. Nelson) E.H. Zacharias	San Joaquin saltbush			1B
Fabaceae (Legume)	*Lathyrus jepsonii* Greene var. *jepsonii*	Delta tule pea		SC	1B
	Astragalus tener A. Gray var. *tener*	Alkali milkvetch			1B
Orobranchaceae (Broomrape)	*Chloropyron molle* (A. Gray) A. Heller subsp. *molle*	Soft salty bird's-beak	FE	R	1B
PLANT TAXA FORMERLY CONSIDERED SPECIAL STATUS:[a]					
Apiaceae (Carrot)	*Lilaeopsis masonii* Mathias & Constance[b]	Mason's lilaeopsis		R	1B
Scrophulariaceae (Figwort)	*Limosella australis* R. Br. (syn. *L. subulata*, naturalized)	Australian (Delta) mudwort			2

NOTE: Listing status: FE = federally listed as endangered and critical habitat designated; R = rare under California Native Plant Protection Act; SC = species of special concern. CNPS (California Native Plant Society) designations: 1B = rare and endangered in California and elsewhere; 2 = rare, threatened, or endangered in California but more common elsewhere.

[a] Recent taxonomic clarifications warrant removal from rare species lists.

[b] See Bone et al. 2011; Fiedler et al. 2011.

integrate vegetation types and key environmental parameters such as landscape position, geomorphology, and biogeochemistry, providing information useful for conservation and restoration management (e.g., Cowardin et al. 1979; Ferren et al. 1996a, 1996b). Our basic units include submerged aquatic vegetation, emergent vegetation of tidal wetlands, natural estuarine-terrestrial ecotones, and emergent vegetation in diked and managed wetlands.

Submerged aquatic vegetation beds feature rooted flowering plants that grow primarily below the water surface in subtidal areas. Linear-leaved pondweeds such as sago pondweed (*Stuckenia pectinata*) and fineleaf pondweed (*S. filiformis*) occur in sloughs and bays and are extensive around islands and other shallow, subtidal areas in Honker Bay, Suisun Cutoff, and Suisun Bay (Baye and Grewell 2011; K. Boyer et al. unpublished data). Widgeon grass (*Ruppia maritima*) and horned-pondweed (*Zannichellia palustris*) also occur in subtidal areas (Mason 1972; Baye and Grewell 2011), thriving in low-salinity, shallow, warm water, although widgeon grass also tolerates salinity fluctuations and mesohaline (5–18 ppt salt concentration) to hyperhaline (>40 ppt) conditions (Brock 1979; Kantrud 1991). Curlyleaf pondweed (*Potamogeton crispus*), an invasive European species, was naturalized in Delaware by 1860 and transported westward in water with fish introductions through the early 20th century (Wehrmeister 1978; Stuckey 1979). It was established in the Delta by the late 1940s (Mason 1957) and, along with native coontail (*Certaophyllum demersum*), has recently spread in the artificially freshened reach of Montezuma Slough and nearby channels in the eastern portion of the Marsh.

Emergent tidal wetlands are regularly inundated by oligohaline (i.e., nearly fresh, 0.5–5.0 ppt) to mesohaline (i.e., 5–18 ppt) tidal flooding regimes driven by daily and monthly tidal waves. Tidal wetland plant communities in the Marsh are diverse and include at least 123 taxa, 73% of which are native, from 34 plant families (Baye and Grewell 2011). Tidal wetland plant communities are best described in terms of two major estuarine geomorphic units that integrate vegetation pattern with topographic and hydrologic features: fringing marshes and tidal marsh platforms (Whitcraft et al. 2011). Marsh platforms are broad, terrace-like landforms that are dissected, and made sinuous and dendritic, by complex tidal channels. Fringing marshes are narrow, linear marshes with simple internal tidal-drainage channel patterns or without drainage channels. Fringing marshes border natural lowlands, hillslopes, or artificial levees. Within each marsh landform, complex vegetation structure develops in relation to internal marsh drainage and elevation patterns. The lower intertidal zones of fringing marshes along sloughs and bay shores are dominated by California bulrush, hardstem bulrush—"tules" (*Schoenoplectus californicus* and *S. acutus* var. *occidentalis*), and cattails (*Typha* spp.). Tufted hairgrass (*Deschampsia cespitosa* subsp. *holciformis*), Lyngbye's sedge (*Carex lyngbyei*), common reed (*Phragmites australis*), and Mason's lilaeopsis (*Lilaeopsis masonii;* but see Bone et al. 2011; Fiedler et al. 2011) are locally common along sloughs. The highest intertidal zones of fringing marshes and tidal marsh plains (i.e., mature platforms) support vegetation flooding only by infrequent spring tides. Rushes, bulrushes, grasses, and salt-tolerant perennial forbs are distributed among higher tidal creek banks and natural levees, marsh plains, ponds, and turf pans. Common species include saltgrass (*Distichlis spicata*), pickleweed (*Sarcocornia pacifica*), Baltic rush (*Juncus arcticus* subsp. *balticus*), three-square bulrush (*Schoenoplectus americanus*), marsh gumplant (*Grindelia stricta* var. *angustifolia*), sea arrow-grass (*Triglochin maritima*), slender arrow-grass (*T. concinna*), fleshy jaumea (*Jaumea carnosa*), and perennial pepperweed (*Lepidium latifolium*). Suisun thistle (*Cirsium hydrophilum* var. *hydrophilum*) and soft bird's-beak (*Chloropyron molle* subsp. *molle*) are endangered plants restricted to emergent tidal wetlands (table 4.1). Poorly drained interior portions of wide Suisun marsh plains (e.g., Rush Ranch Cutoff Slough–Mallard Slough marshes), more remote from drainage of tidal creeks, develop soil conditions with relatively greater influence of waterlogging or persistent shallow flooding, despite relatively high tidal elevation range (Whitcraft et al. 2011). Poorly drained interior marsh plains, now rare in Suisun Marsh because of extensive diking, are often locations of pickleweed- and saltgrass-dominated vegetation where salts concentrate. In climate phases of high rainfall, however, poorly drained interior high-marsh plains and depressions may support large stands of three-square bulrush (Whitcraft et al. 2011).

Estuarine–terrestrial ecotones are transitional areas in the landscape where emergent wetland vegetation intergrades with upland plant species. Here, along the margins of alluvial fans, native perennial and nonnative annual grasses predominate. Scattered within the tidal margins of these ecotones are annual forbs such as smooth goldfields (*Lasthenia glabrata*). The sparse, dwarf vegetation of turf pans includes pickleweed, toad rush (*Juncus bufonius*), alkali weed (*Cressa truxillensis*), and slender arrow-grass (*Triglochin concinna*). Toward the supratidal end of the ecotone, extensive stands of native creeping wildrye (*Leymus triticoides*) are often dominant where cattle have been excluded. These clonal grass swards provide high-functioning wildlife habitat structure and refuges (e.g., high-tide escape cover or waterfowl nesting habitat), unlike areas where native grass has been replaced by alien annual grasses and forbs. Creeping wildrye grows downslope and intergrades with halophytes (i.e., plants that tolerate and grow in soil with high concentrations of salts) such as pickleweed, saltgrass, alkali weed, and alkali heath (*Frankenia salina*). Native field sedge (*Carex praegracilis*) and basket sedge (*C. barbarae*) are clonal species found near seeps or in swales with seasonally saturated or mesic soils, often in association with creeping wild rye.

Diked managed wetlands are the predominant wetland type in today's Suisun

Marsh, and the dikes themselves have a distinct plant community. Dikes enclos-
ing former baylands are maintained at higher elevations than interior ponds and
adjacent waterways. Vegetation on the dikes depends highly on management
actions. Wetland managers impose regular disturbance regimes (e.g., levee capping, grading, mowing, and burning) that promote ruderal vegetation characteristic of regularly disturbed upland habitat. Perennial pepperweed, wild mustard
(*Brassica* spp.), wild radish (*Raphanus sativa*), fennel (*Foeniculum vulgare*), poison hemlock (*Conium maculatum*), Himalayan blackberry (*Rubus armeniacus*),
and other weeds dominate these frequently disturbed areas.

Management of wetland areas within dikes includes a range of activities that
target vegetation types thought to be important to waterfowl (Mall 1969; Rollins
1973; Miller et al. 1975; Suisun Resource Conservation District (SRCD) 2008).
Species composition and growth of wetland vegetation are controlled primarily
by the depth, duration, and frequency of flooding—especially by long periods of
artificial drainage imposed by management—and secondarily by the concentration of salts in the root zone (Mall 1969; Mall and Rollins 1972). Subtle differences
among a range of management plans recommended to Suisun Marsh landowners
(SRCD 2008) favor suites of plant species with different inundation and salinity
tolerances. Landowners may also burn, disk, mow, and/or apply herbicides to
seasonal wetlands in summer to remove "undesirable" vegetation.

Managed wetlands that are flooded for 4 to 5 months, with early drawdown
following waterfowl hunting season, favor production of alien forbs with high
salinity tolerance, such as fathen (*Atriplex prostrata*), brass buttons (*Cotula
coronipfolia*), and western sea-purslane (*Sesuvium verrucosum*). These species
produce abundant seeds eaten by wintering waterfowl in the Marsh (Burns 2003).
Halophytic subshrubs (pickleweed and alkali heath) and graminoid–forb assemblages including saltgrass, Baltic rush, rabbitfoot grass (*Polypogon monspeliensis*), and several other species co-occur in these wetlands. Wetlands flooded for 7
to 8 months, with late drawdowns, support tuberous and nontuberous bulrushes
(e.g., salt marsh bulrush and chairmaker's bulrush) and are managed to force
germination of alkali bulrush from moist soil seed banks (Rollins 1981). Alkali
bulrush (*Bolboschoenus maritimus* subsp. *paludosus*) was mistakenly treated as
Scirpus campestris (misapplied: Jepson 1923–1925) and as *S. robustus* (misapplied:
Mason 1957; George et al. 1965; Mall 1969) for decades in Suisun Marsh and
throughout California, and putative hybrids (*B. maritimus × B. fluviatilis*) have
been noted (Browning et al. 1995; Smith 2002). Alkali bulrush spreads using
rhizomes and has low seed production (Adam 1990), and the seeds often remain
dormant for long periods or have low germination rates (O'Neill 1972; George
and Young 1977).

A small portion of the diked Marsh is irrigated in summer to grow watergrass
(*Echinochloa crus-galli*), a tall summer annual that produces seed consumed

by waterfowl (Miller 1987). Pale smartweed (*Persicaria lapathifolia*) and dotted smartweed (*P. punctata*) often grow with watergrass.

Permanently flooded ponds within diked marshes—like their natural antecedent, historical marsh ponds within tidal marsh plains—support dense stands of sago pondweed. Historically, natural tidal marsh ponds attracted large numbers of canvasbacks and other waterfowl to feed on the nutritious shoots and tubers of the pondweeds (Jepson 1904, 1905; Stoner 1937; Arnold 1996). Some managed wetlands retained permanent ponds following construction of dikes (George et al. 1965; Miller et al. 1975). Other managers have created small permanent ponds within their properties. Both nontidal ponds and tidally choked ponds within Suisun Marsh support widgeon grass and horned pondweed (Mason 1972), as well as sago pondweed (Mason 1957, 1972; Miller et al. 1975). Widgeon grass and sago pondweed are also found in feeder ditches of managed wetlands (Miller et al. 1975). In addition to these submerged plant species, shallow ponds and their edges often support many wetland plant species from diverse functional groups (e.g., free-floating, creeping emergent, tall emergent) that thrive in permanently flooded conditions.

HISTORICAL VEGETATION DYNAMICS: IS THE PAST PROLOGUE?

Holocene History: Paleoecology and Stratigraphy

Suisun Marsh stratigraphy and paleoecology—specifically, depositional records of past responses to long-term fluctuations in salinity regimes under gradual sea-level rise in the late Holocene —provide evidence of past boundary conditions of tidal wetland vegetation in Suisun Marsh. Long-term droughts, prolonged periods of elevated or reduced salinity, extreme floods, unstable mudflat–tidal marsh transitions, and the wide range tidal marsh accretion rates during the late Holocene all are recorded in the stratigraphy of the Marsh sediments. The entire 4,000+ year depositional history of Suisun tidal marshes has recently been investigated by Byrne et al. (2001), Malamud-Roam and Ingram (2004), Malamud-Roam et al. (2007), and Wells et al. (1997). Their investigations deepen the limited historical baseline of 19th-century presettlement conditions, which are now understood to represent an exceptionally benign, cool, moist climate period (Malamud-Roam et al. 2007). The full late Holocene record of marsh prehistory has revised substantially our understanding of the full range of variability expressed in tidal marsh vegetation composition and dynamics in relation to long-term variability in salinity and sediment supply.

Knowledge of Suisun Marsh's prehistoric past is also instructive for prediction of future responses of its vegetation to long-term salinity and sediment-supply trends. Prolonged high-salinity phases were correlated with decreased freshwater inflow from the Delta and resulted in increased abundance of pickleweed and grass species during 1,600–1,300 yr BP, 1,000–800 cal yr BP, 300–200 yr BP, and ca. AD 1950–present. Prolonged low-salinity phases with increased freshwater inflow, congruent with abundant pollen from bulrush and aster family species in sediment cores, occurred before 2,000 yr BP, from 1,300 to 1,200 yr BP, and from ca. 150 yr BP to AD 1950 (Malamud-Roam and Ingram 2004). Climate variations reflected in these findings occurred during a period of relatively slow sea-level rise, which in itself is significant: no stratigraphic record of stable tidal marsh formation or maintenance exists during the relatively rapid sea-level rise that preceded the relatively stationary conditions (slow average sea-level rise, ca. 1.5 mm/yr; Byrne et al. 2001) of the past 750 years.

Stratigraphic and pollen evidence at Rush Ranch indicate that no stable high-marsh platforms or characteristic diverse mature tidal marsh vegetation like those of the 19th and 20th centuries existed before 2,500 yr BP (Byrne et al. 2001). The earliest stages of Suisun Marsh tidal marsh development appear to have consisted of alternating sequences of early-succession tidal marsh and mudflat, interpreted as unstable marsh and mudflat. In the early stage of marsh develop-ment, much of it was prone to periodic reversion to tidal flats and then back to early-succession marsh again. Deep core samples from Southampton Marsh at Benicia also exhibit alternating sequences of estuarine mineral sediment (bay mud), transitions to peat, and peat through their nearly 4,000-year chronology (Malamud-Roam and Ingram 2004). The initial phase of marsh–mudflat alterna-tion in the Marsh occurred during a period of relatively high average salinity, approximately 15–20 ppt, as indicated not by dominance of a tule–cattail assem-blage, but, rather, by the dominance of native California cordgrass (*Spartina foliosa*), a salt marsh pioneer in low tidal marsh (Byrne et al. 2001).

Paleosalinity reconstruction from long sediment cores collected at Rush Ranch also provides a new perspective on the salinity regime of Suisun Marsh at the time of 18th–19th century Euro-American settlement, when Suisun Bay was named "Boca del Puerto Dulce" or "Freshwater Bay" (de Anza 1776; Gudde 1969; Whipple et al. 2012). The period following 750 yr BP was characterized by unusu-ally high freshwater inflows to Suisun Marsh, inferred to be analogous with those of the Sacramento–San Joaquin Delta prior to water diversions in ca. 1930, with fresh to brackish conditions in both winter and summer (Byrne et al. 2001). This was the lowest-salinity phase of the entire Rush Ranch tidal marsh record (Malamud-Roam and Ingram 2004) and, therefore, is not likely to represent a stable modern or future condition. Watson and Byrne (2009) concluded that Suisun Marsh experienced long periods in which either fresh–brackish or salt marsh species were dominant, punctuated by abrupt change.

Reconstructions of past salinities of Suisun Marsh are based on pollen and diatom indices that consider salinity tolerance ranges of plant species and their relative abundance of plant species in stratigraphic layers of core samples. These data clearly indicate that past natural millennial-scale droughts, such as the 1,580-year period 170–1,750 yr BP, caused low freshwater inflows comparable to those caused by 20th-century (post-1930) diversion of Delta outflows. Drought, real or human-made, results in persistent shifts in relative abundance of Marsh vegetation toward salt marsh dominants (pickleweed, saltgrass, and cordgrass), with corresponding reductions of brackish assemblages of sedge-family bulrushes and aster-family forbs (e.g., gumplant, yarrow [*Achillea millefolium*], marsh baccharis [*Baccharis glutinosa*], Suisun aster, and Suisun thistle). Suisun Marsh appears to have originated as bona fide salt marsh and, after 2,000 yr BP, alternated among plant assemblages with different salinity requirements. Thus, Suisun Marsh was never a stable brackish marsh: it alternated among salt marsh, brackish marsh, and fresh–brackish marsh, in cycles lasting from decades to centuries. The transitions among states were driven by climate fluctuations and variable Delta outflows as sea level slowly and steadily increased. Many native plants characteristic of modern brackish Suisun Marsh have shifted locations within local salinity gradients, surviving wide shifts in climate and salinity-driven plant assemblages.

Modern Anthropogenic Influences

An early written description of Suisun Marsh comes from the diary of Captain Juan Bautista de Anza (de Anza 1776), who led a colonizing expedition of California for Spain. De Anza described Native Americans catching fish from boats made of tules in a gulf of fresh water now called Suisun Bay (chapter 2). Large village sites of the Suisun tribe of the Patwin people were in the Suisun area (Kroeber 1925). The Patwin regularly burned tule stands and grasslands at the Estuary's margins to improve the quality of fiber for textile use, remove litter, and open areas of the marsh for waterfowl hunting (Johnson 1978). Stems and rhizomes of emergent macrophytes were harvested for shelter, clothing, boats, basketry, duck decoys, and food (Johnson 1978), likely changing the structure and composition of plant communities. A century following de Anza's expedition, Solano County, which included "92,000 acres of swamp and overflow lands" was settled (Munro-Fraser 1879). In 1876, William Chapman was granted a patent to 12,056 acres on "Grisly Island" and began diking and draining of marshlands for reclamation. Agricultural development of the island then commenced (Frost 1978).

By the 19th century, cattle grazing and crop production resulted in the introduction into tidal wetlands of alien plant species, many of agricultural origin or affinity (e.g., wild celery, beet, and asparagus) as well as several annual European grasses. Also, early settlers of the time recalled that saltgrass was the dominant vegetation on Grizzly Island prior to diking (George et al. 1965). Haying and harvest of saltgrass and three-square (chairmaker's) bulrush as commodities (Mason

1957; George et al. 1965) likely had intensely negative effects on what are now rare, endemic plant species (Whitcraft et al. 2011).

Ultimately, pre–20th century impacts on estuarine vegetation were significant and, coupled with the concomitant urbanization of Solano County through the 20th century, resulted in the loss of wetland acreage and types, including, because of their vulnerable position, transitional plant communities at estuarine margins. Observations from the early 20th century by the well-known botanist and University of California professor Willis Linn Jepson, a Solano County native, provide a colorful retrospective of extant plant communities and a reminder of their aesthetic appeal:

> The whole (Suisun Marsh) region abounds in vegetation. Tall rank tule and rice grasses fringe the sloughs, and in the low lands behind this dense curtain grow all manner of water and marsh plants. Sedges and rushes thrive, wild celery, water hemlock, and water pennywort grow in the wettest places; asters and leather root where the land is a bit higher; while an abounding profusion of sunflowers and sunflower-like plants, twice a man's height, feeding on sunshine, water and fertile soil, run everywhere into a riot of bloom in August and September and at that season color the flat landscape as far as one may see. (Jepson 1905)

Duck Hunting Club Influences on Marsh Flora.   Management for waterfowl in Suisun Marsh has included significant introductions of nonindigenous plant species, as well as nonlocal genotypes of species (from out of state) considered native to the Marsh. Consequently, nonindigenous populations of plant species have persisted in the Marsh's flora. Observations by Jepson (1904, 1905) of ponds in the Ibis Gun Club in the western portion of the Marsh were the first to recognize the importance of sago pondweed as food for waterfowl (Kantrud 1990). Thereafter, aquatic biologists increasingly recognized pondweeds as extremely important plants in the diet of migratory waterfowl (McAtee 1911; Martin and Uhler 1939; Moyle and Hotchkiss 1945) and conducted extensive, original research to support establishment of large sago pondweed stands for wildlife (e.g., McAtee 1917, 1939; Sharp 1939). In Suisun Marsh, Browning found all species of ducks feeding on sago pondweed, with mallard and pintail particularly foraging on it early in the season (see George et al. 1965; Miller et al. 1975). In the 20th century, planting of pondweeds and other aquatic plant species into managed wetlands and lakes was considered good management to improve fish and wildlife habitat; it became a common practice across the country (Stuckey 1979; Hart et al. 2000). For this reason, nonlocal genotypes of native pondweeds and alien species, such as curlyleaf pondweed (*Potamogeton crispus*), were introduced across the United States by wetland managers (Stuckey 1979).

The growing need for intensive management of migratory waterfowl habitat spurred more research on establishment and cultivation of other wildlife food plants (e.g., Kadlec and Wentz 1974). Specialized nurseries have been supplying seeds and tubers of wetland plants to waterfowl habitat managers for nearly a century. Mason (1957) noted that most seed planted by duck clubs in California that had been purchased from out-of-state sources was not genetically adapted to local conditions and provided little improvement to foraging habitat. In 1958, the California Department of Fish and Game (CDFG) and the Soil Conservation Service (SCS) imported wetland plants from around the world, screening more than 500 species for waterfowl food and commercial production of released cultivars (Whitaker 1969; Clary and George 1983). Nonlocal genotypes of smart-weeds (*Persicaria lapathifolia* and *P. pensylvanica*) were field tested at Gray Lodge (Butte County) and then introduced at 10 other wetlands in the Central Valley and Suisun Marsh (Clary and George 1983). Alkali bulrush and many other accessions of wetland plant species were grown in artificial ponds by SCS in Pleasanton, prior to field testing. Alkali bulrush with seed yields as high as 3,500 lbs/acre were identified, but early attempts to establish stands were unsuccessful because of difficulties in handling and germination of seed lots, competitive effects of other plant species, and probably because introduced genotypes were not locally adapted (Miller et al. 1975; George and Young 1977; Clary and George 1983).

Waterfowl food-habit studies by CDFG biologists in Suisun Marsh shifted attention to establishment and seed production of alkali bulrush and other seed plants (George 1963; Mall 1969). Records of seeding rates and acres of intentional plant introductions deemed important by CDFG (e.g., barley [*Hordeum* spp.], watergrass, and alkali bulrush) were inventoried for Suisun Marsh management plans in the 1970s (Miller et al. 1975). Some duck club managers in Suisun Marsh also introduced alien swamp timothy (*Crypsis schoenoides*) from populations in the Sacramento Valley and alien sago pondweed, using tubers mail-ordered from game food plant nurseries, with positive responses from waterfowl (B. Grewell, personal observation). When populations of alien species are introduced and established in wetlands, often other weeds are also unintentionally introduced.

Alien Species Introduction and Dispersal. The legacy of targeted management for alien weeds (e.g., watergrass, fathen, brass buttons, and swamp timothy) in diked wetlands has resulted in dispersal of these species into nearby tidal wetlands. Tidal wetlands and diked wetlands in Suisun Marsh are implicitly components of a single ecosystem, but they are often viewed as discrete habitat types that can be managed separately. Tidal water intimately links these wetlands, however, in large part because flowing water is the primary dispersal mechanism for seeds or clonal ramets of most wetland plants (Nilsson et al. 2010). Additionally, the lack of proactive weed management in tidal wetlands (e.g., for perennial pepperweed) has resulted in rapid reinvasion of alien plants into diked wetlands following control efforts.

The colonization and spread of alien plant species have highly altered wetland plant communities in the Marsh. During the past 20 years, the invasiveness of many weedy species has dramatically increased. Alien invaders competitively displace native species, change the structure and function of wildlife habitat, and alter primary production, food webs, and biogeochemical cycling of nutrients at the ecosystem level. For example, perennial pepperweed and common reed (*Phragmites australis*) are problem species in both managed and tidal wetlands. Perennial pepperweed was established in Yolo and Solano counties by 1941 (Robbins 1941, Robbins et al. 1941) and was present throughout the Bay–Delta and on Grizzly Island by 1960 (Mooney et al. 1986). In the early 1990s, the species spread exponentially along dikes and into diked and tidal wetlands throughout the Marsh (Whitcraft et al. 2011). Today, this aggressive Eurasian mustard threatens the viability and recovery of endangered, endemic Suisun thistle and soft bird's-beak (Fiedler et al. 2007; Grewell et al. 2007).

Common reed is a large, cosmopolitan perennial grass found in wetlands worldwide. Cretaceous fossils and archeological records confirm the native status of *P. australis* in North America (Orson et al. 1987), and it is acknowledged that common reed stands have been part of the Suisun Marsh flora for millennia. Both native and alien subspecies (morphologically distinct; Swearingen and Saltonstall 2010) are now present in Suisun Marsh (Saltonstall 2002). Native *P. australis* subsp. *berlandieri* is typically noninvasive in tidal marshes but, like cattails and tules, it can spread extensively during wet years, owing to its superb flood tolerance and ability to colonize disturbed sediment in both tidal and diked marsh. This native taxon is the host plant of a native butterfly, the Yuma skipper (*Ochlodes yuma*), while the alien *P. australis* subsp. *americanus* hosts the eastern broad-winged skipper (*Poanes biator*) (Shapiro and Manolis 2007). The relative distributions and zones of habitat overlap of the native and alien giant reed taxa in the Marsh are poorly known, however. Current management recommendations call for herbicide control of all common reed stands (SRCD 2008, and tidal restoration plans) and should be reconsidered.

Numerous other alien plant species have invaded emergent tidal wetlands and, if not controlled, will seriously compromise the success of future wetland restoration efforts. Fat-hen is a copious seed producer grown by waterfowl habitat managers, but it is an aggressive secondary invader following removal of perennial pepperweed in tidal wetlands, compromising local endangered-species recovery efforts (B. Grewell, unpublished data). Wild celery (*Apium graveolens*), a garden escapee originally from Europe, has been naturalized "in marshy grounds throughout the Bay Area" since the late 19th century (Greene 1894). This weed recently has spread exponentially in tidal wetlands and directly threatens the recovery of endangered Suisun thistle and soft bird's-beak (Fiedler et al. 2007; Whitcraft et al. 2011). In particular, survivorship of endangered soft bird's beak is compromised when the hemiparasitic plants form physiological connections to the roots of winter annual host plants such as rabbitfoot grass and sickle grasses (*Hainardia cylindrica* and *Parapholis incurva*) that die before the hemiparasites complete their life cycles (Fiedler et al. 2007). Soft bird's-beak survivorship and plant community diversity are both enhanced when perennial native host communities (saltgrass and pickleweed) support soft bird's-beak's parasitic habit (Grewell 2008).

Free-floating and rooted emergent aquatic plants of alien origin have invaded subtidal areas of Suisun Marsh. Floating water hyacinth (*Eichhornia crassipes*) is highly invasive in the Delta and has spread westward to Brown's Island. Rooted, submerged Eurasian watermilfoil (*Myriophyllum spicatum*) and Brazilian waterweed (*Egeria densa*) have invaded nearshore subtidal areas of eastern Suisun Marsh, both of which can hinder movement of fish between subtidal and tidal areas (Simenstad et al. 2000b). Alien weeds often spread during periods when salinity is dampened in estuaries, but then become established and persist when salinity rises. Yellow flag iris (*Iris pseudacorus*) and submerged aquatic beds of alien curlyleaf pondweed were previously restricted to the Delta and to Brown's Island (Grewell 1992), but both have spread aggressively along southeastern Montezuma Slough, where Suisun Marsh salinity control gates operate to artificially dampen salinity regimes (chapter 3). Native Lyngbye's sedge stands recently have expanded in fringing marsh along Suisun tidal sloughs (Baye and Grewell 2011) but may currently be threatened by the spread of yellow flag, which has competitively displaced Lyngbye's sedge stands along tidal sloughs in Oregon (Wilson et al. 2008). The rare California hibiscus (*Hibiscus lasiocarpos* var. *occidentalis*) populations in the south Delta also are threatened by this aggressive iris, and California hibiscus on Brown's Island shorelines may be at risk. In its native range, yellow flag tolerates mesohaline to polyhaline (to 24 ppt) salinity regimes and grows well in freshwater, brackish, and salt marshes (Sutherland 1990; Sutherland and Walton 1990). Although the distribution of yellow flag currently is restricted to oligohaline conditions in Suisun Marsh, this species has high potential for further spread and will likely tolerate the higher-salinity pulses that are expected with sea-level rise and climate change.

CONCEPTUAL FRAMEWORK: RESPONSE OF VEGETATION TO ENVIRONMENTAL CHANGE

The fate of Suisun Marsh vegetation in this century is expected to depend on multiple external drivers (i.e., physical, abiotic forcing factors; see chapter 3), significant feedbacks among them, and cascading biotic–ecological processes (see figure 4.1). These interactions are expected to result in significant changes in Marsh vegetation as critical thresholds of salinity and submergence tolerance are approached, causing rapid changes in vegetation like those that occurred in Suisun Marsh during past episodes of climate change. This should not be surprising, because plants growing in an estuarine marsh face a harsh and highly variable environment and so are often on the threshold of their physiological tolerances (Adam 1990). Phenotypic plasticity may initially lead to rapid evolution of novel traits that allow plants to survive changing conditions and may also increase the invasiveness of alien plant species (Drenovsky et al. 2012). Plant traits with a direct relationship to fitness (e.g., specific leaf area or resource use efficiency) coupled with the outcome of species interactions (e.g., competition or herbivory) across environmental gradients contribute to the structure and composition of marsh plant communities (e.g., Crain et al. 2004; Grewell 2008). Nevertheless, tidal inundation and fluctuating salinity regimes are the primary influences that explain plant distribution and abundance in the present and future Marsh (Hinde 1954; Mahall and Park 1976; Atwater et al. 1979).

In the next sections, we follow the conceptual model in figure 4.1 and describe how Marsh vegetation will likely respond to major changes, especially (1) increases in atmospheric carbon dioxide, (2) decreases in suspended sediment, and (3) sea-level rise (salinity and depth).

Increases in Atmospheric Carbon Dioxide

Anthropogenic increases in carbon dioxide, the major driver of climate change, can have a fertilizing effect on some plants but are expected to differentially affect plant species in tidal wetlands depending on their carbon fixation pathway (see box 4.1).

Suspended Sediment

Feedbacks between sea-level rise, mineral sediment deposition and accretion rates, and marsh vegetation growth strongly influence plant communities in tidal marshes (Redfield 1965; Kirwan and Murray 2007; Mudd et al. 2009, 2010). If the rate of sedimentation is too low to allow buildup of marsh surfaces with rising sea level, many existing plant communities will not survive in place. Some threshold changes in sediment factors that affect vegetation have already occurred in Suisun Marsh and are likely irreversible.

Recent reductions in sediment source and supply to Suisun Marsh have produced suspended sediment regimes in Suisun Marsh that are unprecedented (see chapters 2 and 8). The recent threshold change in reduced suspended sediment throughout San Francisco Estuary has returned the concentrations of suspended sediment in Suisun Marsh to a condition comparable to pre–Gold Rush conditions (Schoellhamer 2011). This reduction of long-term suspended sediment

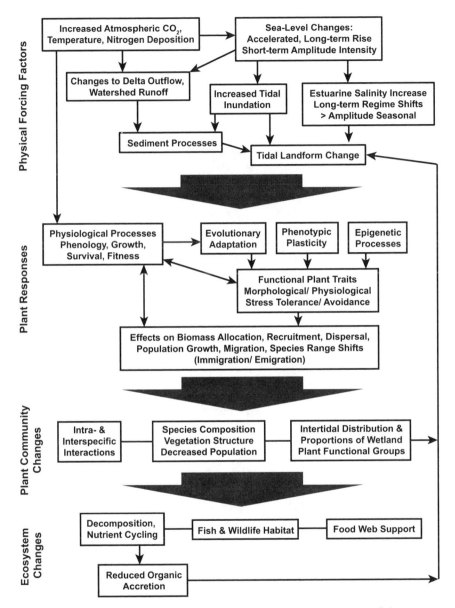

FIGURE 4.1. Conceptual model of effects of climate change on structure of plant communities in Suisun Marsh. Climate-induced changes to interactive, physical forcing factors have cascading effects on wetland plant species, populations, and communities. Ultimately these changes shape the distribution and abundance of estuarine vegetation and supporting tidal landforms.

BOX 4.1. PLANT PHOTOSYNTHESIS AND
RESPONSES TO CLIMATE CHANGE

Atmospheric carbon dioxide (CO_2) is the source of carbon (C) that plants turn into organic compounds (e.g., carbohydrates). This is usually through C_3 or C_4 photosynthesis, whereby light from the sun is absorbed by chlorophyll in plant cells and CO_2 is taken from the air. CO_2 enrichment increases photosynthetic rates and plant growth. Using energy derived from the sun, plants with C_3 photosynthetic pathway assimilate CO_2 into a three-carbon molecule, whereas CO_2 is fixed into four-carbon molecules by "C_4 plants." Globally, most plant species utilize C_3 photosynthesis, but other pathways have evolved in response to a need for increased water-use efficiency, such as in arid or saline environments. Because of their unique photosynthetic pathway, aquatic plants with C_4 photosynthesis (e.g., common reed, cordgrass, saltgrass, and watergrass) are less affected by CO_2 enrichment than C_3 plants (e.g., alkali bulrush, three-square bulrush, California bulrush, cattail, Lyngbye's sedge, and pickleweed), whereas increased CO_2 through increased growth rates alters biomass allocation of C_3 sedges. These fundamental changes in physiological processes change species abundances through shifts in competitive advantage (Rasse 2005). In general, plants that fix carbon by the C_4 pathway lose their competitive advantage in conditions of CO_2 enrichment (Ehleringer and Monson 1993). Long-term experimental studies in Chesapeake Bay have confirmed a compositional shift in dominance from a cordgrass (*Spartina patens*) to a C_3 sedge (three-square bulrush) with CO_2 enrichment, and C_4 plant biomass also declined with rising sea level (Erikson et al. 2007). Native saltgrass and some weed species (common reed and watergrass) are plants with C_4 photosynthesis in Suisun Marsh that may become less competitive with rising CO_2. However, growth of perennial pepperweed (C_3) increases significantly with rising atmospheric CO_2 (Blank and Derner 2004), which may partially explain its increased invasiveness in Suisun Marsh. A change in the relative abundance of C_3 and C_4 plant species has implications for primary productivity, for fauna dependent on marsh plants for food or cover (Ehleringer et al. 1997), and for subsequent changes to ecosystem-level processes (e.g., decomposition and organic sediment accretion). In areas of the Marsh where emergent wetlands are expected to drown and become subtidal habitat, we can predict that the physiological tolerance of plant species to increased water depth with sea-level rise will be more important than relative responses and distribution of C_3 versus C_4 plant species to increasing temperature and atmospheric CO_2. However, along conservation corridors where the Marsh is expected to migrate in response to rising tides, these physiological differences among species may all be extremely important drivers of plant community assembly and persistence. The relative importance of C_3 and C_4 plant species' response to climate change needs investigation as a predictor of plant community shifts.

concentrations is critically important for two vegetation types in Suisun Marsh: submerged aquatic vegetation in subtidal bays, and in sloughs and tidal marshes. Long-term trends of submerged aquatics in oligohaline to mesohaline beds are strongly influenced by turbidity and light attenuation that limit their growth (Orth et al. 2010). Increased light penetration in the water column, particularly during the spring–summer submerged aquatic growing season, is likely to promote colonization and expansion of these beds in shallow subtidal sloughs and bays. Subtidal beds dominated by sago pondweed have expanded dramatically along Suisun Bay shorelines the past 10 years and may continue to expand where conditions of relatively low turbidity, tidal current, and wave energy are suitable for colonization and growth. Sago pondweed appears to have a longer history in Suisun Marsh than the Marsh itself: Goman and Wells (2000) observed pondweed seeds in Suisun sediment cores dating back over 5,000 years, before tidal marsh peats were widely established, and pondweed persisted through the strong salinity fluctuations of the late Holocene. Widgeon grass was prevalent along shallow Suisun Bay shorelines during higher-salinity conditions of the extended drought in the late 1980s (B. Grewell, personal observation). During future phases of prolonged high salinity, the species composition of submerged aquatic beds will likely shift from stands dominated by oligohaline-tolerant sago pondweed that typically dies back when water salinity exceeds 15 ppt (Kantrud 1990) to increased relative abundance of salinity-tolerant widgeon grass with similar overall vegetation structure.

Perhaps the most significant change in Suisun Marsh vegetation will be a shift from emergent macrophytes (*Schoenoplectus acutus, S. californicus,* and *Typha* spp.) to submerged aquatic vegetation in diked marshlands that become permanently flooded following levee failure or engineered tidal restoration. As sea-level rise accelerates, submerged aquatic vegetation beds may become very extensive in formerly diked wetlands if sustained sediment-starved conditions cause subtidal aquatic habitats to persist.

Sea-level Rise

Sea-level rise will adversely affect estuarine wetlands along the entire west coast. Half of all tidal flats and brackish marshes along the Oregon and Washington coasts are projected to be lost to rising seas in the next 90 years (Butler 2003). Future marsh drowning is also predicted to result in the significant loss of marsh in the San Francisco Estuary (Stralberg et al. 2011). Marshes will be lost because sediment accretion rates will not likely be sufficient to counteract sea-level rise. The average long-term rate of marsh accretion revealed by Suisun Marsh sediment cores during the past 750 years has been slow (close to 1.5 mm/yr), corresponding to slow sea-level rise similar to that in the 20th century (Bryne et al. 2001). The accretion rate is significantly less than would be needed for maintaining Marsh vegetation given predicted rates of sea-level rise of 4–6 mm/yr (Kirwan et al. 2010), and substantially less than refined estimates that project a middle range of eustatic sea-level rise of 9 mm/yr (Rahmstorf et al. 2012).

Where marshes are not lost completely, climate change and accelerated sea-level rise are expected to trigger a cascade of major ecological changes, including shifts in plant species distributions (Parker et al. 2011). Suisun Marsh is particularly susceptible to qualitative changes of vegetation states because of its median position in the Estuary's salinity gradient, and because of the greater subsidence of its peaty marsh soils in diked areas compared with those of marshes in San Francisco and San Pablo bays. Two major factors that interact to affect plant communities in the Marsh as sea level rises are salinity and depth.

Salinity

Salinity change appears to have been one of the most influential drivers of change in Marsh tidal vegetation in both historical and prehistoric times (Byrne et al. 2001; Watson and Byrne 2009, 2012) and should be assumed to be a primary agent of future Marsh vegetation change under accelerated sea-level rise and other aspects of climate change (Knowles and Cayan 2002, 2004). The salinity-regime changes that affect tidal marsh vegetation in Suisun include not only average long-term salinity increases, but also the amplitude of variation in seasonal salinity (i.e., extreme summer peak-salinity pulses and winter–spring fresh pulses) and the frequency of extreme, persistent high-salinity events (i.e., prolonged droughts; Callaway et al. 2007). Rapid vegetation change in Suisun Marsh tidal marshes has been observed during prolonged droughts when salt marsh species increase in relative abundance (Atwater et al. 1979; Goman and Wells 2000).

The effect of salinity change on lower intertidal brackish marsh, which is subject to daily tidal flushing, is relatively predictable even without significant changes in mean sea level or tidal range. Future salinity increases will likely reduce the abundance and distribution of tules and cattails in the Marsh's lower intertidal zones and increase the relative abundance of California cordgrass or its hybrids with the alien smooth cordgrass (*S. alterniflora*) (Ayres et al. 2008). Alternatively, these increases could create fluctuating assemblages of alkali bulrush and cordgrass that are currently common in salt–brackish transition zones of North Bay marshes (Pearcy and Ustin 1984; Baye et al. 2000; Diggory and Parker 2011). Alkali bulrush has high salinity tolerance and occurs at a wide range of salinities (0.162–308.0 g/L salts), although growth is typically best at intermediate salinities (Kantrud 1996). The interactive effects of salinity and inundation will eventually override the initial competitive advantage that C_3 cattails, tules, and bulrush gain from increasing atmospheric carbon dioxide (see box 4.1). Complete conversion from tules to low cordgrass salt marsh is particularly likely with a strong interaction between elevated salinity and a rapid sea-level rise that increases submergence periods (Watson 2004; Callaway et al. 2007; Watson and Byrne 2009). Alkali bulrush, which can avoid peak salinity stress through aboveground dieback and dormancy of corms (Pearcy and Ustin 1984, may replace stands of less salinity-tolerant three-square bulrush in poorly drained Suisun tidal marsh plains if summer salinities become too elevated. On the positive side, such elevated salinities may constrain the spread of the noxious perennial pepperweed that is currently so invasive in fresher parts of the Marsh (Watson and Byrne 2009).

The effects of salinity change on marsh plants at higher elevations is complicated by drainage patterns and topography, likely becoming spatially more variable as sea level rises (Watson and Byrne 2009, 2012; Whitcraft et al. 2011). Well-drained tidal areas may potentially undergo only moderate increases in salinity as the degree of tidal flushing increases with increasing submergence frequency, maintaining the tules and similar species. However, soils in poorly drained areas of high tidal marsh platforms, a feature associated with geomorphically mature tidal marshes, may become increasingly salty as a result of evapotranspiration by the plants. As such areas become increasingly salty, dominance may shift to pickleweed, unless tidal drainage increases with sea-level rise (Watson and Byrne 2009). Curiously, survival of pickleweed seedlings may increase with rising sea level, but the height and biomass of mature pickleweed will decrease with increasing salinity (Woo and Takakawa 2012). Salinity and inundation seem to interact to influence net production of pickleweed, but within-site environmental conditions also influence its rise to dominance (Schile et al. 2011). Thus, higher salinity of tidal water strongly reduces production of pickleweed in poorly drained areas but has no effect on the productivity of populations in well-drained conditions (Schile et al. 2011).

Depth

Salt-tolerant marsh plant assemblages will remain in place under sea-level rise only if they can maintain optimal depths for growth through accretion of self-produced organic sediment and through capture of suspended sediment from outside sources. Permanently reduced suspended sediment levels in Suisun Marsh will therefore constrain the ability of tidal marsh areas to accrete sufficient sediment to keep depths shallow enough to favor plant growth (Callaway et al. 2007). Likewise, higher salinity and waterlogging of poorly drained soils in areas with limited development of tidal channels are conducive to vegetation dieback and formation of pans and ponds (Reed and Cahoon 1992; Kirwan and Guntenspergen 2012). These open areas have low sediment-trapping capacity and low organic accretion rates. Extensive pan and pond systems historically developed within interior high-marsh plains such as those near Cordelia (U.S.

Department of Agriculture 1914, 1930), even during the late Holocene, when sea-level rise was comparatively slow.

With longer periods of tidal inundation, well-drained fringing marshes may initially show an increase in tall, dense beds of tules, but these will likely give way to California cordgrass as salinity rises. The biomass of common reed and alkali bulrush is reduced when the plants are inundated at depths greater than 80 cm (Coops et al. 1996). For alkali bulrush, depth tolerance appears to be greatest in fresh water (Kantrud 1996), while biomass is highest when average water depth is 35 cm (Lieffers and Shay 1982).

Sea-level Rise and High Tidal Marsh

Plants of high tidal marsh are, for the most part, least well adapted to the increasing inundation and salinities that accompany sea-level rise. Not only are these species most likely to be stressed by changing conditions, they also occupy areas farthest from sources of mineral sediment accretion that would alleviate the effect (Allen 2000). High-marsh zones in the Marsh occur at two distinct positions: within the Marsh along natural levees, and along the Marsh edges. Plants characteristic of the marsh–terrestrial ecotone can potentially persist by moving upslope (i.e., landward with estuarine transgression). But this can only occur where landward migration in vegetation zones is not constrained by urban development, agriculture, roads, or flood infrastructure. Permanent reduction in suspended sediment in Suisun Marsh and long-term salinity increases may reduce productivity and make the more extensive "internal" high marsh associated with drainage networks vulnerable to submergence and conversion to lower marsh vegetation.

The potential loss of high marsh poses an important conservation dilemma for 21st-century Suisun Marsh (box 4.2). The high-marsh zones in Suisun Marsh support two federally listed rare plants, soft bird's-beak and the endemic Suisun thistle, as well as other rare plants (Fiedler et al. 2007; Grewell et al. 2007; Whitcraft et al. 2011).

Sea-level Rise and Diked Wetlands

Predicting vegetation change in diked areas of Suisun Marsh is complex because diverse factors—including intensive water management, maintenance of levees, the economic and land-use policy contingencies of dike failure and rehabilitation, and sea-level rise—influence what vegetation will be supported. We review two predictable scenarios, both assuming sea-level rise: (a) continued existing land use as managed wetlands behind dikes with restricted hydrology and (b) dike breaching for tidal and subtidal wetland restoration.

Existing management of duck clubs and wildlife areas requires seasonal flooding and gravity drainage and results in marsh subsidence due to peat decom-

BOX 4.2. A COMPARISON OF TWO RARE
WETLAND PLANTS OF SUISUN MARSH

Peggy L. Fiedler, Megan Keever, and Esa Crumb

Suisun Marsh is host to a significant number of rare plant species, including several protected by state and/or federal endangered species legislation. Two species in particular, Suisun thistle and Mason's lilaeopsis, illustrate the management challenges of protecting plant species with different specific habitat preferences within a dynamic wetland ecosystem—challenges made all the more difficult as we learn more about the science of climate change and the notion of rarity.

Suisun thistle (Asteraceae: *Cirsium hydrophilum* var. *hydrophilum*) is an herbaceous, short-lived, perennial thistle restricted to moist or wet habitats of the salt and brackish marshes of the Suisun Marsh ecosystem. Presumed extinct until rediscovered on Grizzly Island in 1989, it was federally listed as endangered in 1997 on the basis of a narrow distribution, low population numbers, and threats to its existence (i.e., altered hydrology, competition from native and nonnative plants, and seed predation by both the thistle weevil [*Rhinocyllus conicus*] and larvae of the butterfly *Phyciodes mylitta*). This highly localized rare plant is considered to be in general decline.

Prior to extensive surveys at Rush Ranch in 2003, Suisun thistle was believed to exist in only a few, small, remnant populations throughout the high-marsh zone of the Marsh ecosystem. These populations occurred at three broad regions, extending across a total area of approximately 1 acre: less than 10 individuals at Grizzly Island Wildlife Area, less than 100 at Peytonia Slough Ecological Reserve, and between 2,000 and 3,000 at Rush Ranch. Surveys conducted by an environmental consulting firm in 2003 at Rush Ranch demonstrated a rare-plant distribution across the 4.3 km² (1,050 acres) of high marsh that far exceeded previous estimates (L.C. Lee and Associates 2003). A total of 47 subpopulations were mapped at the site, only two of which were previously documented, with a total geographic extent of 8.55 acres, all of which belong to a large, single population of approximately 137,500 (22,300–873,200) individuals. Preliminary size-class distribution data suggested that recruitment of new individuals was likely sufficient to maintain this population size.

The total geographic extent of the endangered thistle was expanded from less than 0.005 km² (1.2 acres) across three sites to more than 0.03 km² (74 acres) at Rush Ranch alone. Population size increased from a few thousand plants to an average estimate of 137,500 individuals at Rush Ranch (L.C. Lee and Associates 2003). Importantly, the study revealed that the ditching completed by the State Resources Agency for mosquito abatement provides artificial habitat that may have helped expand the Suisun thistle's geographic distribution within the previous two decades. In summary, the Suisun thistle appeared to be locally abundant, despite being considered extinct a decade ago. However, major threats to population persistence were identified.

Threats to the short- and long-term viability of Suisun thistle observed in the past decade include (1) loss of habitat to invasion by perennial pepperweed, which was associated with 85% of subpopulations; (2) presence of a nonnative, phytophagous, biocontrol weevil (*Rhinocyllus conicus*, capable of reducing seed set by 86% in closely related *Cirsium* species; Turner et al. 1987); (3) habitat destruction by feral alien pigs (*Sus scrofa*; damaged 34% of subpopulations); and (4) potential hybridization with another alien congener, bull thistle (*Cirsium vulgare*, which co-occurred with 45% of subpopulations; L.C. Lee and Associates 2003; Fiedler et al. 2007). These threats, in addition to rising sea levels that will surely negatively affect the first-order channel habitats of this rare species, should be expected to diminish the distribution and abundance of Suisun thistle throughout the Marsh.

Mason's lilaeopsis (Apiaceae: *Lilaeopsis masonii*) is a California state-listed rare species known throughout San Francisco Bay–Delta region, including the tidally influenced main stem of the Napa River. Because of the relative inaccessibility of this rarity's intertidal habitat, the biology and geographic distribution of Mason's lilaeopsis was poorly known for nearly a half century, until mitigation funding provided an opportunity to examine this species along with

FIGURE 4.2. (Left) Suisun Marsh thistle (*Cirsium hydrophilum* var. *hydrophilum*) capitulescence (photo by P. Baye). (Right) Flowering ramets of Mason's lilaeopsis (*Lilaeopsis masonii*; photo by M. Keever).

other rare plants of the intertidal zone of the Suisun Marsh and Sacramento–San Joaquin Delta.

A four-year study documented all populations of Mason's lilaeopsis throughout the 182 km² (44,973 acres) of its known range and characterized the plant's habitat. Field surveys conducted by boat, canoe, and on foot revealed that, unlike the Suisun thistle, Mason's lilaeopsis is one of the most widespread "rare" species in California (Golden and Fiedler 1991). Mason's lilaeopsis reproduces readily by seed and by fragmentation, colonizing, and persisting on an exceptionally wide range of substrates and soil types, from riprap rock to sand. Another significant finding at that time was the profound morphological variation between the diminutive Mason's lilaeopsis and a widespread, robust congener, western lilaeopsis (*L. occidentalis*). This extensive variation confounds reliable taxonomic identification, particularly for those specimens intermediate between the two species (one protected by state law and one not), a condition known as the "Goldilocks conundrum" (Fiedler et al. 2011).

Several years later, several of these same researchers investigated the genetic basis of this morphological variation to determine whether Mason's lilaeopsis is sufficiently distinct from its closely related, widespread congener to continue to warrant specific status (and rare conservation dollars for its protection). Fiedler et al. (2011) examined two portions of the *Lilaeopsis* genome in seven species and, through DNA sequence analysis of ITS1, 5.8S, and ITS2 nuclear ribosomal DNA, revealed no differences between Mason's lilaeopsis and western lilaeopsis. Further genetic analysis using a combination of fragment data from three AFLP primers reflected very minor differences among all samples, with a single cluster of Mason's lilaeopsis + western lilaeopsis samples unequivocally distinct from the five other *Lilaeopsis* species. Bone et al. (2011) further corroborated these findings through additional genetic analyses, including of the chloroplast DNA genome.

This brief discussion of two rare plant species of Suisun Marsh serves to illustrate the belief that rare plants are idiosyncratic in their causes of rarity (Fiedler 1986). Suisun thistle and Mason's lilaeopsis both are rare species whose distributions are strongly influenced by twice daily tidal inundation within the same marsh landscape. One species, Suisun thistle, is intrinsically rare, although a focused survey greatly expanded the known geographic range and population size. However, this thistle remains threatened by anthropogenically derived stressors. The other species, Mason's lilaeopsis, may well be "rare" only in a legal sense, as it is exceptionally geographically widespread for any species considered rare. Further, it faces few threats of any kind, reproduces vigorously (sexually and asexually), and is capable of expanding its range by taking advantage of the complex hydrodynamics of the Marsh and Delta ecosystems. Such comparisons illustrate that management of rare species, particularly those most vulnerable to rising sea levels, requires individualistic approaches using the best available science.

position (chapter 3). Sea-level rise will reduce the number of hours per tidal cycle during which gravity drainage of subsided diked marsh can occur. Either a shift to active pumping to drain the marshes or increased residual ponding of water will result. As the salinity of tidal intake water for duck clubs increases, poor drainage and residual ponding will cause increased residual salinity and greater production of soil sulfides during flooding and soil saturation. We can also expect increased acid sulfates and iron oxides during drainage periods. Few plant species can tolerate extremes of acid sulfate and hypersalinity; brass buttons and fathen are relatively tolerant of acid sulfate soils in seasonal saline ponds, but pickleweed would likely decline under "red salt pan" pond conditions (i.e., where iron oxide crusts and acid sulfates concentrate).

Alternatively, reduced feasibility of levee maintenance and pumping may provide economic and ecological incentives for conversion to traditional water management regimes of "circulating ponds" with choked tidal flows and some levee overtopping (George et al. 1965). Vegetation under this management regime would consist of deeply flooded sago pondweed beds and fringing tule marsh that grade to alkali bulrush stands. As sea level rises, increased levee maintenance that includes sediment capping to raise and eventually widen levees will be necessary to maintain existing, nontidal, flood-drain water management in these areas. This will likely cause increasingly frequent vegetation disturbance and increased ruderal alien vegetation on levees (e.g., perennial pepperweed, radish, and poison hemlock). Also as sea level rises, a reduction of native perennial forbs and shrubs with limited intrinsic dispersal and inundation tolerances can be expected along current brackish marsh edges. In the final diked-marsh scenario, accelerated sea-level rise may overstep levees and cause cascading levee failures. This scenario would be similar to conversion of the wetlands to extensive subtidal aquatic vegetation; its likelihood would depend on how much sea level rises and the return interval of extreme (El Niño) high sea levels and storms (chapters 3 and 9). The fate of vegetation in tidally breached subsided diked wetlands is not necessarily what has been intended for "restoration" of brackish tidal marsh as expected in 20th-century conditions.

Under future accelerated sea-level rise, increased salinity, and persistent suspended-sediment conditions, submerged aquatic vegetation beds are more likely to rapidly colonize and dominate breached subtidal basins and undergo mud accretion rates lower than those before 1999. Mudflat–tidal marsh succession may be preempted under such conditions, depending on local sediment concentration, so submerged aquatics may represent an alternative stable state for tidally breached baylands in Suisun Marsh. Other interactions between sea-level rise and lower suspended-sediment concentrations in the Marsh may affect channel bank retreat and tidal marsh edge erosion, including slough widening, and associated vegetation processes and structure. Increased tidal prism and reduced suspended sediment in channel systems within peaty tidal marshes may result in excess energy that induces slumping and scour of banks.

Exposure of active marsh scarps and overhanging banks, and restriction of tall emergent low-marsh vegetation by erosion, may facilitate some high-marsh plants subject to light competition from tall tule canopies. Similarly, small, fugitive colonial species of erosional banks such as Australian (Delta) mudwort (*Limosella australis*), water pennywort (*Hydrocotyle verticillata*), and Mason's lilaeopsis may be supplied expanded microhabitat by chronic bank erosion. Increased marsh scarp erosion and retreat bordering open bays and wide sloughs is also likely to result from increased wave energy associated with sea-level rise and tidal prism increase. The horizontal extent of tidal marsh loss due to increased wave energy may reverse trends of 20th-century marsh progradation (particularly giant reed marsh) in northern Suisun Bay.

PREPARING FOR THE FUTURE: WHAT CAN WE DO?

Successful adaptive management frameworks to prepare for future environmental change in the Suisun Marsh require long-term planning. Some wetland plant species can be preserved if they are able to disperse and migrate inland along with the marshes, as wetlands have moved with changes in the sea through geologic time. However, few areas that can support colonization remain, so protecting these areas for marsh expansion may reduce the overall loss of estuarine vegetation. Planning should include land acquisition, tidal marsh restoration, and protection of adjacent lowland areas. In the face of climate change, wetland plant communities are responding to a complex set of factors. Prudence suggests that the following five actions will likely be important for conservation of the wetland flora of Suisun Marsh.

Conserve undeveloped lowland plains and valleys with low-intensity agriculture to accommodate sea-level rise at least 3 m above the modern high tide line. It is highly uncertain whether peat or mineral sediment accretion rates of Suisun tidal marsh platforms will be able to keep pace with accelerated sea-level rise, particularly under persistent low suspended-sediment regimes, reduced productivity of more frequently flooded marsh, and higher salinity (chapter 3). The process of tidal marsh formation by submergence of adjacent areas is relatively independent of marsh sediment accretion rates and, thus, is relatively resilient even with accelerated sea-level rise. This is particularly true of the species-rich high marsh and the terrestrial–marsh ecotone, which together support most of the rare endemic or endangered plant species of Suisun Marsh. Within the western San Francisco Estuary, only Suisun Marsh retains extensive, undeveloped, gently sloping valleys, hollows, flats, and alluvial fans bordering tidal and diked wetlands. Conserving them for future passive estuarine expansion is a regionally unique opportunity, and a necessity to ensure that at least minimal tidal marsh areas will persist even at higher levels of forecast sea-level rise rates. Passive estuarine transgression is not only less risky than primary reliance on tidal sedimentation or peat accretion for tidal marsh maintenance at higher rates of sea-level rise, but it would entail less restoration engineering and cost, and would not depend on engineered sediment placement.

Conserve freshwater seasonal streams and seeps bordering the Marsh. As the Estuary's salinity gradient shifts upstream, brackish marsh gradients within Suisun Marsh will likely contract around local surface-stream discharge zones and localized groundwater discharge zones at the terrestrial edges, as we now see in San Pablo and San Francisco bays. These brackish gradients may become refuges for the prevalent, typical vegetation assemblages of Suisun tidal marsh.

Make high brackish marsh vegetation a high conservation and restoration priority. High brackish marsh supports the greatest plant species diversity and most of the rare and endangered species of Suisun Marsh, terrestrial vertebrates as well as plants. It is also the vegetation type most prone to conversion by drowning or shifting to lower marsh assemblages during accelerated sea-level rise, especially in higher-elevation areas and tidal creek banks. High marsh on terrestrial ecotones is capable of upslope migration as sea level rises (figure 4.3). Land acquisition for tidal marsh conservation and restoration designs should prioritize conservation or construction of gently sloping terrestrial edges to accommodate persistent high brackish marsh, particularly near freshwater discharge zones.

Implement comprehensive weed management as a component of wetland conservation and restoration efforts. As marshlands migrate upslope with rising tides and colonization of conservation buffer areas proceeds, a "first come, first served" approach must be avoided because rapidly assembled plant communities will favor aggressive colonizers, usually highly competitive invasive alien species, that will establish themselves on marsh edges. Integrated weed management strategies should minimize nontarget effects and simultaneously address multiple potentially invasive species to avoid inadvertent secondary invasions ("musical chairs syndrome"; Smith et al. 2006) inherent in current single-weed management efforts in the Suisun Marsh.

Improve understanding of significant limiting factors for plant species and community adaptation to climate change. As plant species shift their distribution, plant community assembly will depend on the ability of individual plant species to persist and disperse; on demographic processes; and on interactions among

FIGURE 4.3. Ephemeral vernal annual assemblage of *Lasthenia glabrata* and *Triphysaria versicolor* in tidal alluvial fan ecotone of southern Rush Ranch (photo by P. Baye).

new combinations of plants, native and alien. Research on variation in key functional traits of plant species (e.g., inundation tolerance, salinity tolerance, and demography), populations (dispersal and colonization patterns), and communities (species interactions) in response to novel combinations of environmental conditions is needed to improve our understanding of wetland plant communities' responses to climate change. It may then be possible to link variation in key traits at these multiple levels of ecological organization to spatial, dynamic process-based models, which can be used to refine conceptual models and conservation management plans (Pressey et al. 2007; Traill et al. 2011).

Prepare integrated conservation management strategies that facilitate dispersal, assisted colonization, and persistence of rare plants. Rare or uncommon plant species in Suisun Marsh typically lack well-connected habitats and distributions within efficient dispersal distances among (usually fragmented and small) remnant populations. As marsh drowning and vegetation type conversions become increasingly frequent with sea-level rise and salinity increases, remnant populations of rare species are likely to become stranded beyond colonization reach of suitable habitat during estuarine transgression. Facilitated dispersal consists of transplanting or seeding populations into appropriate, possibly assembled, plant communities in a new range that is favorable for persistence of the populations under future conditions. This will not be a simply horticultural exercise. Integrated conservation strategies that include management for habitat connectivity, maintenance of physical processes, consideration of conservation genetics, and (if warranted) assisted colonization of species that are not able to shift their ranges (see Vitt et al. 2010; Loss et al. 2011) may be needed to facilitate the colonization, adaptation, and persistence of rare or increasingly uncommon tidal marsh plants. Assisted colonization is recommended as an important component of long-term conservation strategies for tidal marsh and biodiversity in Suisun Marsh in a changing climate.

Manage the Marsh as an interconnected, rapidly changing, novel ecosystem.   It is essential to recognize the interdependence between diked managed wetlands and tidal wetlands in the Marsh. Management of vegetation in these hydrologically linked components of the Suisun Marsh waterscape as independent systems is outdated, simplistic, and unrealistic. Wetland managers need to develop the perspective that Suisun Marsh will never be returned to presettlement conditions, that long-term sustainability of critical habitats will require a hydrogeomorphic approach to restoration, and that restoration of natural hydrologic processes may mitigate some of the effects of sea-level rise (Euliss et al. 2008; Smith et al. 2008; Brown and Humburg 2010). A move away from targeted management for invasive plant species in diked wetlands and implementation of active management of invasive weeds in tidal wetlands will improve habitat support functions for rare as well as common species. Given projections for increases in subtidal habitat, a shift to more emphasis on native pondweeds may return to importance in the Suisun Marsh. Features that will enhance habitat for migratory birds should be identified and incorporated in tidal restoration and buffer corridor projects.

CONCLUSIONS

Wetland vegetation varies considerably along the estuarine continuum from San Francisco Bay to the Delta. The estuarine flora of Suisun Marsh is distinctive and supports a number of rare and endangered plant species. We anticipate the need to rethink current restoration and land management strategies to address future conservation needs of endemic and endangered plant species and viable estuarine plant communities in the Marsh. Conservation and management of Suisun Marsh flora has never been as challenging as it is in today's highly modified environment, with looming large-scale changes in physical processes due to climate change compounding that challenge. Integrated assessments that consider specifically how climate change will modify physical processes and biota in the Estuary are beginning to emerge and are framed in the context of multiple drivers of change (see figure 4.1).

Long-term planning with emphasis on providing buffers for the future movement of plants inland may, overall, reduce the loss of estuarine vegetation and should include the prioritization of land acquisition, tidal marsh restoration, and protection of adjacent upland areas where corridors for wetland migration with sea-level rise are possible. Slightly brackish, oligohaline conditions are maintained artificially in Suisun Marsh to support waterfowl habitat management, on the assumption of a causal relationship between managed low-salinity regimes and vegetation that supports high waterfowl habitat quality. The evidence for a strong causal relationship between artificially reduced salinity and beneficial vegetation change, however, is lacking, and the impacts of this static waterfowl-habitat management approach on the greater Suisun Marsh ecosystem are poorly known. The success of long-term plans for preservation of Suisun Marsh under climate-change scenarios potentially will be enhanced by a move away from the status quo to an integrated management approach that simultaneously considers estuarine vegetation in tidal and diked wetlands.

REFERENCES

Adam, P. 1990. Saltmarsh Ecology. New York: Cambridge University Press.

Allen, J. R. L. 2000. Morphodynamics of Holocene salt marshes: a review sketch from the Atlantic and southern North Sea coasts of Europe. Quaternary Science Reviews 19: 1155–1231.

Arnold, A. 1996. Suisun Marsh History: Hunting and Saving a Wetland. Marina, California: Monterey Pacific.

Atwater, B. F., S. G. Conard, J. N. Dowden, C. W. Hedel, R. L. MacDonald, and W. Savage. 1979. History, landforms, and vegetation of the Estuary's tidal marshes. Pages 347–385 in San Francisco Bay: The Urbanized Estuary (T. J. Conomos, A. E. Leviton, and M. Berson, eds.). San Francisco: AAAS Pacific Division.

Ayres, D. R., K. Zaremba, C. M. Sloop, and D. R. Strong. 2008. Sexual reproduction of cordgrass hybrids (Spartina foliosa × alterniflora) invading tidal marshes in San Francisco Bay. Diversity and Distributions 14: 187–195.

Baye, P. R., P. M. Faber, and B. J. Grewell. 2000. Tidal marsh plants of the San Francisco Estuary. Pages 9–33 in Baylands Ecosystem Species and Community Profiles: Life Histories and Environmental Requirements of Key Plants, Fish and Wildlife (P. R. Olofson, ed.). Prepared by the San Francisco Bay Area Wetlands Ecosystem Goals Project. San Francisco Bay Regional Water Quality Control Board, Oakland, California.

Baye, P. R., and B. J. Grewell. 2011. Partial flora of estuarine vegetation at Rush Ranch, Suisun Marsh, Solano County, California: vascular plant species. Appendix in Estuarine Vegetation at Rush Ranch Open Space Preserve, San Francisco Bay National

Estuarine Research Reserve, California (C. Whitcraft, B.J. Grewell, and P.R. Baye, authors). San Francisco Estuary and Watershed Science 9(3).

Blank, R.R., and J. Derner. 2004. Effects of CO_2 enrichment on plant–soil relationships of *Lepidium latifolium*. Plant and Soil 262: 159–167.

Bone, T.S., S.R. Downie, J.M. Affolter, and K. Spalik. 2011. A phylogenetic and biogeographic study of the genus *Lilaeopsis* (Apiaceae Tribe Oenantheae). Systematic Botany 36: 789–805.

Brock, M.A. 1979. Accumulation of proline in a submerged aquatic halophyte, *Ruppia* L. Oecologia 51: 217–219.

Brown, D.M., and D.D. Humburg. 2010. Confronting the challenges of climate change for waterfowl and wetlands. Memphis, Tennessee: Ducks Unlimited.

Browning, J., K.D. Gordon-Gray, and S.G. Smith. 1995. Achene structure and taxonomy of North American *Bolboschoenus* (Cyperaceae). Brittonia 47: 433–445.

Burns, E.G. 2003. An analysis of food habitat of green-winged teal, northern pintail, and mallards wintering in the Suisun Marsh to develop guidelines for food plant management. MS thesis, University of California, Davis.

Butler, C.J. 2003. The disproportionate effect of global warming on the arrival dates of short-distance migratory birds in North America. Ibis 145: 484–495.

Byrne, R., L. Ingram, S. Starratt, F. Malamud-Roam, J.N. Collins, and M.E. Conrad. 2001 Carbon-Isotope, diatom, and pollen evidence for late Holocene salinity change in a brackish marsh in the San Francisco Estuary. Quaternary Research 55: 66–76.

Callaway, J.C., V.T. Parker, M.C. Vasey, and L.M. Schile 2007. Emerging issues for the restoration of tidal marsh ecosystems in the context of predicted climate change. Madroño 54: 234–248.

Cayan, D.R., P.D. Bromirski, K. Hayhoe, M. Tyree, M.D. Dettinger, and R.E. Flick. 2008a. Climate change projections of sea level extremes along the California coast. Climatic Change 87 (Supplement 1): S57–S73.

Cayan, D.R., M.D. Maurer, M.D. Dettinger, M. Tyree, and K.M. Hayhoe. 2008b. Climate change scenarios for the California region. Climatic Change 87 (Supplement 1): S21–S42.

Clary, R.F., and H.A. George. 1983. Ten years of testing for waterfowl food plants in California. CAL-NEVA Wildlife Society Transactions 1983: 91–96.

Cloern, J.E., N. Knowles, L.R. Brown, D. Cayan, M.D. Dittinger, T.L. Morgan, D.H. Schoellhamer, M.T. Stacey, M. van der Wegen, R.W. Wagner, and A.D. Jassby. 2011. Projected evolution of California's bay-delta-river system in a century of climate change. PLoS ONE 6(9): e24465.

Coops, H., F.W.B. van den Brink, and G. van der Velde. 1996. Growth and morphological responses of four helophyte species in an experimental water depth gradient. Aquatic Botany 54: 11–24.

Cowardin, L.M., V. Carter, F.C. Golet, and E.T. LaRoe. 1979. Classification of wetlands and deepwater habitats of the United States. U.S. Fish and Wildlife Service FWS/OBS-79-31.

Crain, C.M., B.R. Silliman, S.L. Bertness, and M.D. Bertness. 2004. Physical and biotic drivers of plant distribution across estuarine salinity gradients. Ecology 85: 2539–2549.

de Anza, J. B. 1776. Diary of Juan Bautista de Anza October 23, 1775—June 1, 1776. http://anza.uoregon.edu/anza76.html.

Diggory, Z. E., and V. T. Parker. 2011. Seed supply and revegetation dynamics at restored tidal marshes, Napa River, California. Restoration Ecology 19: 121–130.

Drenovsky, R. E., B. J. Grewell, C. M. D'Antonio, J. L. Funk, J. J. James, N. Molinari, I. M. Parker, and C. L. Richards. 2012. A functional trait perspective on plant invasion. Annals of Botany 110: 141–153.

Ehleringer, J. R., T. E. Cerling, and B. R. Helliker. 1997. C4 photosynthesis, atmospheric CO_2, and climate. Oecologia 112: 285–299.

Ehleringer, J. R., and R. K. Monson. 1993. Evolutionary and ecological aspects of photosynthetic pathway variation. Annual Reviews in Ecology and Systematics 24: 411–439.

Enright, C., and S. D. Culberson. 2009. Salinity trends, variability, and control in the northern reach of theSan Francisco Estuary. San Francisco Estuary and Watershed Science 7(2): article 3.

Erikson, J. E., J. P. Megonigal, G. Peresta, and B. G. Drake. 2007. Salinity and sea level rise mediate elevated CO_2 effects on C_3-C_4 plant interactions and tissue N in a Chesapeake Bay tidal wetland. Global Change Biology 13: 202–215.

Euliss, N. H., L. M. Smith, D. A. Wilcox, and B. A. Browne. 2008. Linking ecosystem processes with wetland management goals: charting a course for a sustainable future. Wetlands 28: 553–562.

Ferren, W. R., Jr., P. L. Fiedler, and R. A. Leidy. 1996a. Wetlands of California, part I: history of wetland habitat classification. Madroño 43: 105–124.

Ferren, W. R., Jr., P. L. Fiedler, R. A. Leidy, K. D. Lafferty, and L. A. K. Mertes. 1996b. Wetlands of California, part II: classification and description of wetlands of the central and southern California coast and coastal watersheds. Madroño 43: 125–182.

Fiedler, P. L. 1986. Concepts of rarity in vascular plant species, with special reference to the genus Calochortus Pursh (Liliaceae). Taxon 35: 502–518.

Fiedler, P. L., E. K. Crumb, and A. K. Knox. 2011. Reconsideration of the taxonomic status of Mason's lilaeopsis—a state-protected rare species in California. Madroño 58: 131–144.

Fiedler, P. L., M. E. Keever, B. J. Grewell, and D. J. Partridge. 2007. Rare plants in the Golden Gate Estuary (California): the relationship between scale and understanding. Australian Journal of Botany 55: 206–220.

Frost, J. 1978. A Pictorial History of Grizzly Island. San Francisco: Trade Pressroom.

George, H. A. 1963. Planting alkali bulrush for waterfowl food. Game Management Leaflet 9. California Department of Fish and Game, Sacramento.

George, H. A., W. Anderson, and H. McKinnie. 1965. An evaluation of the Suisun Marsh as a waterfowl area. Administrative Report 20. California Department of Fish and Game, Sacramento.

George, H. A., and J. A. Young. 1977. Germination of alkali bulrush seed. Journal of Wildlife Management 41: 790–793.

Golden, M., and P. L. Fiedler. 1991. Final Report of the Habitat for Lilaeopsis masonii (Umbelliferae), A California State-Listed Rare Plant Species. Report submitted June 3,

1991, to the Endangered Plant Program, Natural Heritage Division, California Department of Fish and Game, Sacramento.

Goman, M., and L. Wells. 2000. Trends in river flow of the last 7000 yr affecting the northeastern reach of San Francisco Bay Estuary. Quaternary Research 54: 206–217.

Greene, E. L. 1894. Manual of the Botany of the Region of San Francisco Bay. San Francisco: Cubery.

Grewell, B. J. 1992. Vascular plant species observed at Brown's Island, Suisun Bay, Contra Costa County, California. Environmental Services Office Suisun Marsh Progam File Report. California Department of Water Resources, Sacramento.

Grewell, B. J. 2008. Hemiparasites generate environmental heterogeneity and enhance species coexistence in salt marshes. Ecological Applications 18: 1297–1306.

Grewell, B. J., J. C. Callaway, and W. R. Ferren, Jr. 2007. Estuarine Wetlands. Pages 124–154 in Terrestrial Vegetation of California (M. G. Barbour, T. Keeler-Wolf, and A. Schoenherr, eds.). Berkeley: University of California Press.

Gudde, E. G. 1969. 1000 California Place Names: Their Origin and Meaning. Berkeley: University of California Press.

Hart, S., M. Klepinger, H. Wandell, D. Garling, and L. Wolfson. 2000. Integrated pest management for nuisance exotics in Michigan inland lakes. Water Quality Series WQ-56. Michigan State University Extension, East Lansing.

Hickson, D., and T. Keeler-Wolf. 2007. Vegetation and land use classification and map of the Sacramento—San Joaquin River Delta. California Department of Fish and Game, Sacramento.

Hinde, H. P. 1954. The vertical distribution of salt marsh phanerogams in relation to tide levels. Ecological Monographs 24: 209–225.

Jepson, W. L. 1904. Willis L. Jepson Field Notes October 13, 1004. Field Books of Willis L. Jepson 12: 36–47. University and Jepson Herbaria Archives. Berkeley: University of California Press.

Jepson, W. L. 1905. Where ducks dine. Sunset Magazine (February): 409–411.

Jepson, W. L. 1923–1925. A Manual of the Flowering Plants of California. Berkeley: University of California Press.

Johnson, P. J. 1978. Patwin. Pages 350–359 in Handbook of North American Indians, vol. 8: California (R. F. Heizer, ed.). Washington, D.C.: Smithsonian Institution.

Kadlec, J., and W. A. Wentz. 1974. State-of-the-art survey and evaluation of marsh plant establishment techniques: induced and natural, vol. 1. Contract Report D-74-9, U.S. Army Corps of Engineers.

Kantrud, H. A. 1990. Sago pondweed (*Potamogeton pectinatus* L.): a literature review. Resource Publication 176. U.S. Fish and Wildlife Service, Washington, D.C.

Kantrud, H. A. 1991. Wigeongrass (*Ruppia maritima*): a literature review. Fish and Wildlife Research 10. U.S. Fish and Wildlife Service, Washington, D.C.

Kantrud, H. A. 1996. The alkali (*Scirpus maritimus* L.) and saltmarsh (*S. robustus* Pursh) bulrushes: A literature review. Information and Technology Report 6, National Biological Service. http://www.npwrc.usgs.gov/resource/plants/bulrush/index.htm (Version 16JUL97).

Keeler-Wolf, T., M. Vaghti, and A. Kilgore. 2000. Vegetation mapping of Suisun Marsh,

Solano County: a report to the California Department of Water Resources. Unpublished administrative report on file at Wildlife and Habitat Data Analysis Branch, California. California Department of Fish and Game, Sacramento.

Kirwan, M. L., and G. R. Guntenspergen. 2012. Feedbacks between inundation, root production, and shoot growth in a rapidly submerging brackish marsh. Journal of Ecology 100:764–770.

Kirwan, M. L., G. R. Guntenspergen, A. D'Alpaos, J. T. Morris, S. M. Mudd, and S. Temmerman. 2010. Limits on the adaptability of coastal marshes to rising sea level. Geophysical Research Letters 37: L23401.

Kirwan, M. L., and A. B. Murray. 2007. A coupled geomorphic and ecological model of tidal marsh evolution. Proceedings of the National Academy of Sciences USA 104: 6118–6122.

Knowles, N. 2002. Natural and management influences on freshwater inflows and salinity in the San Francisco Estuary at monthly to interannual scales. Water Resources Research 38: 25–1 to 25–11.

Knowles, N., and D. Cayan. 2002. Potential effects of global warming on the Sacramento/San Joaquin watershed and the San Francisco Estuary. Geophysical Research Letters 29: 18–21.

Knowles, N., and D. Cayan. 2004. Elevational dependence of projected hydrologic changes in the San Francisco Estuary and Watershed. Climatic Change 62: 319–336.

Kroeber, A. L. 1925. Handbook of the Indians of California. Bulletin No. 78. Bureau of American Ethnology, Washington, D.C.

L.C. Lee and Associates. 2003. Geographic distribution and population paramenters of the endangered Suisun thistle (*Cirsium hydrophilum* var. *hydrophilum*) at Rush Ranch in Solano County, California. Consultant's report prepared for the Solano Water Agency, December 2003.

Lieffers, V. J., and J. M. Shay. 1982. Distribution and variation in growth of *Scirpus maritimus* var. *paludosus* on the Canadian prairies. Canadian Journal of Botany 60: 1938–1949.

Loss, S. R., L. A. Terwilliger, and A. C. Peterson. 2011. Assisted colonization: integrating conservation strategies in the face of climate change. Biological Conservation 144: 92–100.

Mahall, B. E., and R. B. Park. 1976. The ecotone between *Spartina foliosa* Trin and *Salicornia virginica* L. in salt marshes of northern San Francisco Bay II. Soil water and salinity. Journal of Ecology 64: 793–809.

Malamud-Roam, F., D. Dettinger, B. L. Ingram, M. K. Hughes, and J. L. Florsheim. 2007. Holocene climates and connections between the San Francisco Bay Estuary and its watershed: a review. San Francisco Estuary and Watershed Science 5(1): article 3.

Malamud-Roam, F., and B. L. Ingram. 2004. Late Holocene $\delta^{13}C$ and pollen records of paleosalinity from tidal marshes in the San Francisco Bay estuary, California. Quaternary Research 62: 134–145.

Mall, R. E. 1969. Soil-water-salt relationships of waterfowl food plants in the Suisun Marsh of California. Wildlife Bulletin No. 1. California Department of Fish and Game, Sacramento.

Mall, R. E., and G. Rollins. 1972. Wildlife resource requirements: waterfowl and the Suisun Marsh. Pages 60–68 *in* Ecological Studies of the Sacramento–San Joaquin Estuary (J. E. Skinner, ed.). Delta Fish and Wildlife Protection Study Report No. 8. California Department of Fish and Game, Sacramento.

Martin, A. C., and F. M. Uhler. 1939. Food of game ducks in the United States and Canada. U.S. Department of Agriculture Technical Bulletin No. 634.

Mason, H. L. 1957. A Flora of the Marshes of California. Berkeley: University of California Press.

Mason, H. L. 1972. Floristics of the Suisun Marsh. Appendix B *in* An Environmental Inventory of the North San Francisco Bay—Stockton Ship Channel Area of California, part I: Point Edith to Stockton Area (C. L. Newcombe and H. L. Mason, eds.). Richmond, California: Point San Pablo Laboratory, San Francisco Bay Marine Research Center.

McAtee, W. L. 1911. Three important wild-duck foods. U.S. Bureau of Biological Survey Circular No. 81.

McAtee, W. L. 1917. Propagation of wild-duck foods. U.S. Department of Agriculture Bulletin No. 465.

McAtee, W. L. 1939. Wildfowl Food Plants: Their Value, Propagation and Management. Ames, Iowa: Collegiate Press.

Miller, A. W., R. S. Miller, H. C. Cohen, and R. F. Schultze. 1975. Suisun Marsh Study, Solano County, California. U.S. Department of Agriculture Soil Conservation Service, Washington, D.C.

Miller, M. R. 1987. Fall and winter foods of northern pintails in the Sacramento Valley, California. Journal of Wildlife Management 51: 403–412.

Mooney, H. A., S. P. Hamburg, and J. A. Drake. 1986. The invasions of plants and animals into California. Pages 250–274 *in* Ecology of Biological Invasions of North America and Hawaii (H. A. Mooney and J. A. Drake, eds.). Ecological Studies, vol. 58. New York: Springer-Verlag.

Moyle, J. B., and N. Hotchkiss. 1945. The aquatic and marsh vegetation of Minnesota and its value to waterfowl. Minnesota Department of Conservation Technical Bulletin No. 3.

Moyle, P. B., W. A. Bennett, W. E. Fleenor, and J. R. Lund. 2010. Habitat variability and complexity in the upper San Francisco Estuary. San Francisco Estuary and Watershed Science 8(3): 1–24. http://escholarship.org/uc/item/0kf0d32x.

Mudd, S. M., A. D'Alpaso, and J. T. Morris. 2010. How does vegetation affect sedimentation on tidal marshes: investigating particle capture and hydrodynamic controls on biologically-mediated sedimentation. Journal of Geophysical Research 115 (F3). doi:10.1029/2009JF001566.

Mudd, S. M., S. M. Howell, and J. T. Morris. 2009. Impact of dynamic feedbacks between sedimentation, sea-level rise, and biomass production on near surface marsh stratigraphy and carbon accumulation. Estuarine, Coastal and Shelf Science 82:377–389.

Munro-Fraser, J. P. 1879. History of Solano County. San Francisco: Wood, Alley. [1994 reprint. Fairfield, California: James Stevenson.]

Nilsson, C., R.L. Brown, R. Jansson, and D.M. Merritt. 2010. The role of hydrochory in structuring riparian and wetland vegetation. Biological Reviews 85: 837–858.

O'Neill, E.J. 1972. Alkali bulrush seed germination and culture. Journal of Wildlife Management 36: 649–652.

Orson, R., R.S. Warren, and W.A. Niering. 1987. Development of a southern New England drowned valley tidal marsh. Estuaries 10: 6–27.

Orth, R.J., M.R. Williams, S.R. Marion, D.J. Wilcox, T.J.B. Marruthers, K.A. Mooer, W.M. Kemp, M.R. Dennison, P. Bergstrom, and R.A. Batuik. 2010. Long-term trends in submersed aquatic vegetation (SAV) in Chesapeake Bay, USA related to water quality. Estuaries and Coasts 33: 1114–1163.

Parker, V.T., J.C. Callaway, L.M. Schile, M.C. Vasey, and E.R. Herbert. 2011. Climate change and San Francisco Bay–Delta tidal wetlands. San Francisco Estuary and Watershed Science 9(3): 1–15.

Pearcy, R.W., and S.L. Ustin. 1984. Effects of salinity on growth and photosynthesis of three California tidal marsh species. Oecologia 62: 68–73.

Pressey, R.L., M. Cabeza, M.E. Watts, R.M. Cowling, and K.A. Wilson. 2007. Conservation planning in a changing world. Trends in Ecology and Evolution 22: 583–592.

Rahmstorf, S., G. Foster, and A. Cazenave. 2012. Comparing climate projections to observations up to 2011. Environmental Research Letters 7: 1–5.

Rasse, D.P., G. Peresta, and B.G. Drake. 2005. Seventeen years of elevated CO_2 in a Chesapeake Bay wetland: sustained but contrasting responses of plant growth and CO_2 uptake. Global Change Biology 11: 369–377.

Redfield, A.C. 1965. Ontogeny of a salt marsh estuary. Science 147: 50–55.

Reed, D.J., and D.R. Cahoon. 1992. The relationship between marsh surface topography, hydroperiod, and growth of Spartina alterniflora in a deteriorating Louisiana salt marsh. Journal of Coastal Research 8: 77–81.

Robbins, W.W. 1941. Alien plants growing without cultivation in California. California Agriculture Experimental Station Bulletin No. 637.

Robbins, W.W., M.K. Bellue, and W.S. Ball. 1941. Weeds of California. California State Department of Agriculture, Sacramento.

Rollins, G.L. 1973. Relationships between soil salinity and the salinity of applied water in the Suisun Marsh of California. California Fish and Game 59: 5–35.

Rollins, G.L. 1981. A Guide to Waterfowl Habitat Management in Suisun Marsh. California Department of Fish and Game, Sacramento.

Saltonstall, K. 2002. Cryptic invasion by a non-native genotype of the common reed, Phragmites australis, into North America. Proceeding National Academy of Sciences USA 99: 2445–2449.

San Francisco Estuary Institute. 1998. EcoAtlas Baylands Maps. Richmond, California: San Francisco Estuary Institute.

Schile, L.M., J.C. Callaway, V.T. Parker, and M.C. Vasey. 2011. Salinity and inundation influence productivity of the halophytic plant Sarcocornia pacifica. Wetlands 31: 1165–1174.

Schoellhamer, D.H. 2011. Sudden clearing of estuarine waters upon crossing the threshold from transport to supply regulation of sediment transport as an erodible sediment pool is depleted: San Francisco Bay, 1999. Estuaries and Coasts 34: 885–899.

Shapiro, A. M., and T. D. Manolis. 2007. Field Guide to Butterflies of the San Francisco Bay and Sacramento Valley Regions. Berkeley: University of California Press.

Sharp, W. M. 1939. Propagation of *Potamogeton* and *Sagittaria* from seeds. Transactions of North American Wildlife Conference 4: 351–358.

Simenstad, C. A., S. B. Brandt, A. Chalmers, R. Dame, L. A. Deegan, R. Hodson, and E. D. Houde. 2000a. Habitat–biotic interactions. Pages 427–455 *in* Estuarine Science: A Synthetic Approach to Research and Practice (J. E. Hobbie, ed.). Washington, D.C.: Island Press.

Simenstad, C. A., J. Toft, H. Higgins, J. Cordell, M. Orr, P. Williams, L. Grimaldo, Z. Hymanson, and D. Reed. 2000b. Sacramento–San Joaquin Delta breached levee wetland study (BREACH). Wetland Ecosystem Team, School of Fisheries, University of Washington.

Smith, L. M., N. H. Euliss, D. A. Wilcox, and M. M. Brinson. 2008. Application of a geomorphic and temporal perspective to wetland management in North America. Wetlands 28: 563–577.

Smith, R. G., B. D. Maxwell, F. D. Menalled, and L. J. Rew. 2006. Lessons from agriculture may improve the management of invasive plants in wildland systems. Frontiers in Ecology and the Environment 4: 428–434.

Smith, S. G. 2002. *Bolboschoenus* (Ascherson) Palla in: Flora of North America Editorial Committee, eds. 1993+. Flora of North America North of Mexico 23: 37–44.

Stoner, E. A. 1937. A record of twenty-five years of wildfowl shooting on the Suisun Marsh, California. Condor 39: 242–248.

Stralberg, D., M. Brennan, J. C. Callaway, J. K. Wood, L. M. Schile, D. Jongsomjit, M. Kelly, V. T. Parker, and S. Crooks. 2011. Evaluating tidal marsh sustainability in the face of sea level rise: a hybrid modeling approach applied to San Francisco Bay. PLoS ONE 6: e27388.

Stuckey, R. L. 1979. Distributional history of *Potamogeton crispus* (curly pondweed) in North America. Bartonia 46: 22–42.

Suisun Resource Conservation District. 2008. Adaptive Habitat Management Template. Suisun Resource Conservation District, Suisun City, California.

Sutherland, W. J. 1990. Biological flora of the British Isles: *Iris pseudacorus* L. Journal of Ecology 78: 833–848.

Sutherland, W. J., and D. Walton. 1990. The changes in morphology and demography of *Iris pseudacorus* L. at different heights on a saltmarsh. Functional Ecology 4: 655–659.

Swearingen, J., and K. Saltonstall. 2010. Phragmites field guide: distinguishing native and exotic forms of common reed (*Phragmites australis*) in the United States. Plant conservation alliance, weeds gone wild. http://www.nps.gov/plants/ alien/pubs/index.htm.

Traill, L. W., K. Perhans, C. E. Lovelock, A. Prohaska, S. McFallan, J. R. Rhodes, and K. A. Wilson. 2011. Managing for change: wetland transitions under sea-level rise and outcomes for threatened species. Diversity and Distributions 17: 1225–1233.

Turner, C. E., R. W. Pemberton, and S. S. Rosenthal. 1987. Host utilization of native *Cir-*

sium thistles (Asteraceae) by the introduced weevil *Rhinocyllus conicus* (Coleoptera: Curculionidae) in California. Environmental Entomology 16(1): 111–115.

U.S. Department of Agriculture. 1914. Advance sheet: soil map, San Francisco Bay region sheet, California. *In* Reconnaissance Soil Survey of the San Francisco Bay region California (1917; L.C. Holmes and J.W. Nelson, authors). U.S. Department of Agriculture Field Operations Bureau of Soils and University of California Agricultural Experiment Station.

U.S. Department of Agriculture. 1930. Soil map Suisun area California. *In* Soil Survey of the Suisun Area, California (E.J. Carpenter and S.W. Cosby, authors). U.S. Department of Agriculture Bureau of Chemistry and Soils and University of California Agricultural Experiment Station.

Vitt, P., K. Havens, A.T. Kramer, D. Sollenberger, and E. Yates. 2010. Assisted migration of plants: changes in latitudes, changes in attitudes. Biological Conservation 143: 18–27.

Watson, E.B. 2004. Changing elevation, accretion, and tidal marsh plant assemblages in a south San Francisco Bay tidal marsh. Estuaries 27: 684–698.

Watson, E.B., and R. Byrne. 2009. Abundance and diversity of tidal marsh plants along the salinity gradient of the San Francisco Estuary: implications for global change ecology. Plant Ecology 205: 113–128.

Watson, E.B., and R. Byrne. 2012. Recent (1975–2004) vegetation change in the San Francisco Estuary, California, tidal marshes. Journal of Coastal Research 28: 51–63.

Wehrmeister, J.R. 1978. An ecological life history of the pondweed *Potamogeton crispus* L. in North America. CLEAR Technical Report 99. The Ohio State University Center for Lake Erie Area Research, Columbus, Ohio.

Wells, L.E., M. Goman, and R. Byrne. 1997. Long term variability of fresh water flow into the San Francisco Estuary using paleoclimatic methods. University of California Water Resources Center Technical Completion Report W-834. http://escholarship.org/uc/item/8r366ont.

Whipple, A.A., R.M. Grossinger, D. Rankin, B. Stanford, and R.A. Askevold. 2012. Sacramento–San Joaquin Delta historical ecological investigation: exploring pattern and process. San Francisco Estuary Institute Publication No. 627. Richmond, CA: San Francisco Estuary Institute–Aquatic Science Center.

Whitaker, E.A. 1969. The SCS in wildlife management. Transactions of the California-Nevada Section of The Wildlife Society 5: 72–80.

Whitcraft, C., B.J. Grewell, and P.R. Baye. 2011. Estuarine vegetation at Rush Ranch Open Space Preserve, San Francisco Bay National Estuarine Research Reserve, California. San Francisco Estuary and Watershed Science 9(3).

Wilson, B.L., R.E. Brainerd, B. Newhouse, and N. Otting. 2008. Field Guide to the Sedges of the Pacific Northwest. Corvallis: Oregon State University Press.

Woo, I., and J.Y. Takekawa. 2012. Will inundation and salinity levels associated with projected sea level rise reduce the survival, growth, and reproductive capacity of *Sarcocornia pacifica* (pickleweed)? Aquatic Botany 102: 8–14.

5

Waterfowl Ecology
and Management

Joshua T. Ackerman, Mark P. Herzog, Gregory S. Yarris,
Michael L. Casazza, Edward Burns, and John M. Eadie

Suisun Marsh has long been a favored place for waterfowl (ducks, geese, and swans). Before the first duck clubs were established in 1879, market hunters had used the Marsh for at least 20 years, and they continued to hunt it until market hunting was outlawed in 1918 with the passage of the Migratory Bird Treaty Act. Suisun Marsh's proximity to San Francisco allowed market hunters to get their vast harvest to market relatively quickly by boat. During a single hunting season in 1911–12, an estimated 250,000 ducks were sold in San Francisco markets from all sources, and 350,000 ducks were sold statewide (Garone 2011).

The west side of Suisun Marsh had the first duck clubs, apparently for both the quality of hunting and the ease of access. In 1879, the Southern Pacific Company completed a railroad through the west side of the Marsh from Benicia to Suisun City, but it required constant and costly maintenance because it sank repeatedly into the Marsh (Arnold 1996; Garone 2011). Most sport duck hunters in those days were wealthy, because it was costly both to travel to the Marsh from San Francisco and to lease hunting rights. Decisions about railroad placement and continued maintenance may have been at least partly due to the influence of wealthy duck hunters (Arnold 1996). Access became easier after the 1913 opening of the railroad along the east side of the Marsh, and the establishment of duck clubs followed (Arnold 1996). From 1880 to 1930, with the arrival of railroads and construction of levees to restrain tidal water flow, much of Suisun Marsh was converted to agriculture. However, the success of agriculture was relatively short lived, because increasing soil salinity, at least partially a result of upstream water diversions to irrigate farms in the Central Valley, made agriculture unprofitable.

Ultimately, much of the recently reclaimed farmland was converted back into wetland habitats for duck hunting clubs.

Today, Suisun Marsh is one of the largest contiguous brackish marshes in the western United States and contains 12% of the remaining natural wetlands in California. The continuity and abundance of wetlands in the Marsh is attributable to efforts by duck hunters interested in maintaining high-quality waterfowl habitat. Today, 75% is privately owned and managed for waterfowl habitat and hunting opportunities, usually at considerable expense to the landowners (Gill and Buckman 1974). The majority of these privately held wetlands are managed in consultation with the Suisun Resource Conservation District. Presently, the 158 privately owned duck hunting clubs in Suisun Marsh, together with the California Department of Fish and Wildlife's Grizzly Island Wildlife Area, provide over 21,000 ha of wetland habitats.[1] These diked wetlands are managed to control the daily tidal influence, reduce salt accumulation in the soil, and promote food production for waterfowl. In particular, the Marsh provides important wetland resources during early winter migration (September–November), when many waterfowl have arrived in California but when other wetlands are not yet flooded in the Central Valley. Suisun Marsh also provides waterfowl habitats during drought periods, when other Central Valley wetlands are limited.

OVERVIEW: WINTERING WATERFOWL

Population Trends

California, and in particular the Central Valley, is a major wintering area for waterfowl in North America. Nearly 5 million ducks, geese, and swans winter within California, accounting for 68% of the wintering waterfowl in the Pacific Flyway. Importantly, 41% of the North American breeding population of northern pintail (see table 5.1 for scientific names of birds) winters in California.[2] Suisun Marsh plays an important role in relation to its size (470 km^2 or 116,000 acres), supporting more than 60,000 wintering waterfowl each year (table 5.1).

Wintering waterfowl populations have been surveyed annually in California since 1953 by the California Department of Fish and Game (now Department of Fish and Wildlife) in conjunction with the U.S. Fish and Wildlife Service (USFWS). We used these midwinter waterfowl population indices[3] to assess long-

1. http://www.water.ca.gov/suisun/

2. Based on mean midwinter waterfowl population size indices, 2000–09.

3. Midwinter waterfowl data were collected by the California Department of Fish and Game and U.S. Fish and Wildlife Service and were kindly provided by Dan Yparraguirre, Shaun Oldenburger, Mike Wolder, and Cheryl Strong. The midwinter waterfowl index represents an estimated population abundance at the time and location of each survey and is not extrapolated to estimate true

term trends (since the 1950s) in California as a whole and in Suisun Marsh in particular.[4] Midwinter waterfowl surveys are subject to substantial survey biases when viewed at small temporal and spatial scales, but they provide the only long-term estimates of overall population trends and species composition over the past 60 years. These results represent population indices rather than actual population size estimates. Prior to this period, waterfowl abundance data are limited. Duck hunting clubs' harvest records are among the better sources of data for waterfowl in the late 1800s and early 1900s, and published records are available from several duck hunting clubs in Suisun Marsh.

The Ibis Gun Club (established 1879) and Tule Belle Shooting Club (established 1886) are among the oldest duck clubs in Suisun Marsh (Arnold 1996; Hall 2011). The Tule Belle Shooting Club was characterized as having shallow ponds (more suitable for dabbling ducks), and the Ibis Gun Club had deeper ponds (more suitable for diving ducks; Stoner 1937). Stoner (1934; for Tule Belle Shooting Club) and Stoner (1937; for Ibis Gun Club) obtained club hunting archives and published their harvest records. In the late 19th century, there was no regulatory limit in California on the number of ducks that could be harvested, but after 1901 the limit was 50 ducks per day (Arnold 1996).

During 25 hunting seasons from 1882 to 1907, Ibis Gun Club members (3–5 people per day) harvested 36,126 ducks, 595 geese (mostly snow geese), and 51 swans, with members averaging 20 birds harvested per person per day (Stoner 1937). During 16 hunting seasons from 1885 to 1901, Tule Belle Shooting Club members (4–12 people per day) harvested 20,844 ducks, 441 geese (mostly snow geese), and 8 swans, with members averaging a daily bag of 16 birds (Stoner 1934).

Combining the two duck hunting clubs' records shows that the majority of ducks harvested in the late 1800s (N = 56,970) consisted of pintail (28%), wigeon (25%), green-winged teal (17%; includes some cinnamon teal that were not differentiated), canvasback (15%), shoveler (5%), mallard (5%), scaup and ring-necked ducks (2%; not differentiated), ruddy ducks (1%), bufflehead (1%), and gadwall (less than 1%; Stoner 1934, 1937). Additionally, hunters harvested 1,036 geese (85% snow geese, 10% white-fronted geese, and 5% Canada geese) and 59 swans, likely tundra swans, at the two clubs (Stoner 1934, 1937). Although this is just a sampling of hunter harvest in those days, these two clubs represent waterfowl

---

population sizes. Because additional transects were periodically flown over Suisun Bay, Grizzly Bay, and Honker Bay by the U.S. Fish and Wildlife Service, especially in more recent decades, we removed these counts from the Suisun Marsh totals to eliminate possible bias. We also removed the California Department of Fish and Game's count for scaup in 1955, because it was >121 times the mean scaup count for 1953–62 and likely consisted of data collected from Suisun Bay (the original data sheets did not differentiate between Suisun Marsh and Suisun Bay, unlike latter surveys).

4. The Suisun Marsh survey strata include all wetlands within Suisun Marsh but exclude Grizzly Bay, Suisun Bay, and Honker Bay, where most diving ducks and sea ducks would occur.

TABLE 5.1. Main waterfowl species present in Suisun Marsh.

Their primary habitat, timing of use of Suisun Marsh, past and current population size index for Suisun Marsh and California, and long-term (1953–2011) and short-term (1990–2011) population trends for each species in Suisun Marsh, California, and proportional trend in Suisun Marsh in relation to the statewide population size index. Waterfowl population indices are based on annual midwinter waterfowl surveys conducted by plane. Surveys were not conducted in Suisun Marsh in 2010 or 2011 because of inclement weather. Data are from U.S. Fish and Wildlife Service and California Department of Fish and Wildlife

| | | | Use of Suisun Marsh | |
Species (Family/Subfamily/Tribe)	Scientific name	Main habitat	October–March	April–September
DABBLING DUCKS (ANATIDAE/ANATINAE/ANATINI)				
Northern shoveler	*Anas clypeata*	Freshwater wetlands	X	X
Northern pintail	*Anas acuta*	Freshwater wetlands	X	X
American wigeonb	*Anas americana*	Freshwater wetlands	X	
American green-winged teal	*Anas crecca*	Freshwater wetlands	X	
Mallard	*Anas platyrhynchos*	Freshwater wetlands	X	X
Gadwall	*Anas strepera*	Freshwater wetlands	X	X
Cinnamon teal	*Anas cyanoptera*	Freshwater wetlands	X	X
PERCHING DUCKS (ANATIDAE/ANATINAE/CAIRININI)				
Wood duck	*Aix sponsa*	Riparian, freshwater wetlands	X	X
STIFF-TAILED DUCKS (ANATIDAE/ANATINAE/OXYURINI)				
Ruddy duck	*Oxyura jamaicensis*	Freshwater wetlands	X	X
DIVING DUCKS (ANATIDAE/ANATINAE/AYTHYINI)				
Scaupb	*Aythya affinis and Aythya marila*	Open bay, wetlands	X	
Canvasback	*Aythya valisineria*	Open bay, wetlands	X	
Ring-necked duck	*Aythya collaris*	Freshwater wetlands	X	
Redhead	*Aythya americana*	Open bay, wetlands	X	X
Sea ducks (Anatidae/ Anatinae/Mergini)				
Bufflehead	*Bucephala albeola*	Open bay, wetlands, lakes	X	
Common goldeneyeb	*Bucephala clangula*	Open bay, lakes	X	
Common merganserb	*Mergus merganser*	Rivers, lakes	X	
Surf scoterb	*Melanitta perspicillata*	Open bay, coastal ocean	X	

Midwinter waterfowl index				Winter waterfowl population trends[a]					
(Mean: 1953–62)		(Mean: 2000–09)		1953–2011			1990–2011		
Suisun Marsh	California	Suisun Marsh	California	Suisun Marsh	California	Proportion in Suisun Marsh vs. California	Suisun Marsh	California	Proportion in Suisun Marsh vs. California
21,262	171,576	17,196	532,612	≈	+	≈	≈	+	≈
235,845	1,770,961	14,061	1,210,806	−	−	−	−	+	−
31,962	710,976	7,472	523,962	−	−	≈	≈	≈	≈
11,679	163,597	7,296	405,451	≈	+	−	−	+	−
22,530	541,384	6,402	341,705	−	−	≈	−	≈	≈
2,014	20,168	2,034	166,881	≈	+	−	≈	+	≈
105	1,454	55	6,888	≈	+	≈	≈	≈	≈
0	1,550	0	1,686	na	≈	na	na	≈	na
2,269	37,108	1,439	85,834	≈	+	≈	+	+	+
495	81,316	1,025	111,307	≈	+	≈	≈	≈	+
3,768	55,557	657	50,987	≈	≈	≈	≈	+	≈
44	648	105	57,970	≈	+	≈	≈	+	≈
116	589	17	1,629	≈	+	≈	≈	≈	≈
17	2,512	552	25,750	+	+	+	≈	≈	+
20	2,522	84	2,998	≈	≈	+	≈	≈	≈
0	335	1	5,875	≈	+	≈	≈	≈	≈
0	13,308	0	30,604	≈	+	≈	≈	-	≈

(continued)

TABLE 5.1. *(continued)*

Species (Family/Subfamily/Tribe)	Latin	Main habitat	Use of Suisun Marsh October– March	April– September
American coot	*Fulica americana*	Freshwater wetlands	X	
Taxon groups				
Dabbling ducks[c]	*Anas and Aix*	Freshwater wetlands	X	X
Dabbling ducks (no pintail)[c,d]	*Anas and Aix*	Freshwater wetlands	X	X
Diving ducks[e]	*Aytha and Oxyura*	Open bay, wetlands	X	
Sea ducks	*Melanitta, Bucephala, and Mergus*	Open bay, coastal ocean, lakes	X	
Dark geese[f]	*Branta and Anser*	Agricultural fields, freshwater wetlands	X	X
White geese[f]	*Chen*	Agricultural fields, freshwater wetlands	X	
Swans	*Cygnus*	Freshwater wetlands	X	
Total waterfowl	*Anseriformes*	All of the above	X	X

[a] Waterfowl trends were analyzed using linear regression with alpha = 0.05. Key: + population is significantly increasing, – population is significantly declining.
[b] Some species are not differentiated during midwinter waterfowl surveys, and we have listed the predominant species only. However, Canada goose includes lesser, Taverner, dusky, and western canada goose; cackling goose includes Aleutian and small cackling goose; American wigeon may include some Eurasian wigeon; scaup includes both lesser scaup and greater scaup; surf scoter may include some black scoter and white-winged scoter; common goldeneye may include some Barrow's goldeneye; and common merganser may include some hooded merganser and red-breasted merganser.

abundance at both shallow and deep-water habitats over two decades. Therefore, these data likely characterize the relative species composition of the Marsh, except possibly for ruddy ducks (which hunters likely avoided), and are suggestive of very large population sizes of ducks and snow geese in Suisun Marsh during the late 1800s.

Suisun Marsh continued to be an important wintering area for waterfowl, and state agencies established several public waterfowl areas in the mid-1900s, including Joice Island Refuge in 1931 and Grizzly Island Waterfowl Management Area (now called "Grizzly Island Wildlife Area") in 1952. The continued abundance of ducks was demonstrated on opening day of the hunting season in 1952 at the Grizzly Island Wildlife Area when 598 public hunters harvested 5,200 ducks

Midwinter waterfowl index				Winter waterfowl population trends[a]					
(Mean: 1953–62)		(Mean: 2000–09)		1953–2011			1990–2011		
Suisun Marsh	California	Suisun Marsh	California	Suisun Marsh	California	Proportion in Suisun Marsh vs. California	Suisun Marsh	California	Proportion in Suisun Marsh vs. California
57,375	445,528	12,810	438,216	–	≈	–	≈	+	≈
325,397	3,381,666	54,515	3,189,992	–	–	–	–	+	≈
89,552	1,610,705	40,454	1,979,186	–	≈	≈	≈	+	≈
6,643	167,085	3,243	307,728	≈	+	≈	≈	+	≈
38	18,677	637	65,228	+	+	+	≈	–	+
27,252	307,852	1,490	371,169	–	≈	–	≈	+	≈
8,117	354,278	0	564,321	–	+	–	≈	+	≈
281	15,940	364	73,771	≈	+	≈	≈	+	≈
367,728	4,245,498	60,249	4,572,209	na	na	na	na	na	na

[c] Wood ducks were included in dabbling duck taxa group.
[d] All dabbling ducks except northern pintail.
[e] Ruddy ducks were included in diving duck taxon group.
[f] The dark geese taxon group includes all Canada geese races (*Branta* spp.) and white-fronted goose. The white geese taxon group includes snow goose and Ross' goose.

(9 ducks per hunter; Hall 2007). Since then, 12 additional former duck hunting clubs, as well as Joice Island Refuge, have been added to the Grizzly Island Wildlife Area complex, making it the largest single (6,070 ha) managed property in the Marsh.

Suisun Marsh currently provides wintering habitat for more than 60,000 waterfowl, nearly 13,000 coots, and countless other shorebirds, seabirds, wading birds, and terrestrial birds (table 5.1). Dabbling ducks are the most numerous by far (55,000), followed by diving ducks (3,000), geese (1,500), sea ducks (600), and swans (350).[5] These numbers are well below the Central Valley Joint Venture (CVJV) Implementation Plan's peak population objective of 300,000 ducks win-

5. Mean midwinter waterfowl population size index, 2000–09.

tering in the Suisun Basin (CVJV 2006), and well below the average of 367,700 waterfowl present in Suisun Marsh during the 1950s.[6]

Population trends in the Marsh have declined for dabbling ducks and geese, remained stable for diving ducks and the very few swans that occur there, and slightly increased for sea ducks (figure 5.1). The same trends occur even after accounting for California-wide population trends (figure 5.2), indicating that proportionately fewer dabbling ducks and geese winter in Suisun Marsh than did historically (figure 5.3). Much of the decline in dabbling ducks occurred between 1950 and 1970 and is the result of a significant decline in pintail during this period. Pintail accounted for the majority of dabbling ducks historically (table 5.2); therefore, after removing pintail from our analysis, the data still indicate a slight decline in other dabbling ducks using Suisun Marsh (figures 5.1B and 5.3B). This decline in other dabbling ducks, however, has been much less pronounced since 1980. Also of note, the decline in "dark" geese (Canada geese and white-fronted geese) likely is attributable, in part, to distributional shifts in their wintering areas, rather than population declines or specific avoidance of the Marsh. In fact, small cackling geese and white-fronted geese have experienced population growth in the Pacific Flyway over the past several decades (Collins et al. 2011). Small cackling geese began "short-stopping" in Oregon and Washington instead of continuing their southern winter migration into California. Similarly, the decline in white-fronted geese is partly due to increased flooded rice and wetland habitat within the Sacramento Valley and a corresponding distributional shift of white-fronted geese out of the Suisun Marsh and Sacramento–San Joaquin Delta region in the 1990s (Ackerman et al. 2006).

Many species of dabbling ducks that winter in Suisun Marsh have declined in numbers (figure 5.4). One exception is gadwall, whose population in the Marsh decreased from the 1950s to 1970s, increased in the 1980s, and has been stable since the 1990s (figure 5.4F). Concurrently, gadwall populations have rapidly increased both within California (table 5.1) and continent wide (USFWS 2011). When the statewide increase in gadwall is accounted for, the proportion of California gadwall wintering in Suisun Marsh has actually declined slightly. Green-winged teal have enjoyed population increases in California (table 5.1) and continent wide (USFWS 2011), yet have declined in the Marsh in relation to their statewide abundance. Similarly, shoveler have increased dramatically in California (table 5.1) and continent wide (USFWS 2011), yet their population index in Suisun Marsh has merely remained stable. Therefore, wintering dabbling ducks, by and large, have shown significant declines in the Marsh. One contributing factor to the decline of dabbling ducks in the Marsh compared with other parts of the state in recent decades may be the increased wetland area in

6. Mean midwinter waterfowl population size index, 1953–62.

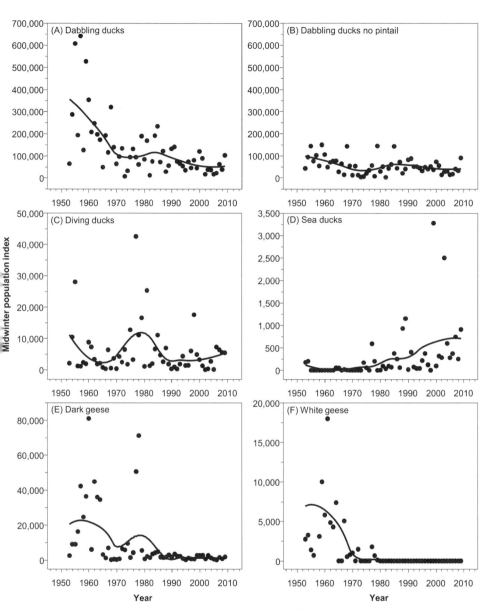

FIGURE 5.1. Midwinter population size indices for waterfowl taxon groups in Suisun Marsh, 1953–2009. Waterfowl are categorized as (A) dabbling ducks, (B) dabbling ducks not including pintail, (C) diving ducks, (D) sea ducks, (E) dark geese, and (F) white geese. See table 5.1 for species within each taxonomic group. Solid line illustrates the population trend and was estimated with a LOESS function. Midwinter waterfowl data were collected by the California Department of Fish and Wildlife and U.S. Fish and Wildlife Service. Suisun Marsh was not surveyed as part of the midwinter waterfowl survey in 2010 and 2011.

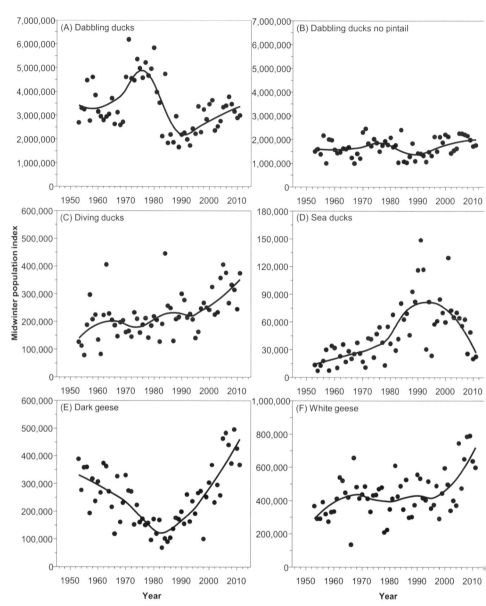

FIGURE 5.2. Midwinter population size indices for waterfowl taxon groups statewide in California, 1953–2011. Waterfowl are categorized as (A) dabbling ducks, (B) dabbling ducks not including pintail, (C) diving ducks, (D) sea ducks, (E) dark geese, and (F) white geese. See table 5.1 for species within each taxon group. Solid line illustrates the population trend and was estimated with a LOESS function (Neter et al. 1996). Midwinter waterfowl data were collected by the California Department of Fish and Wildlife and U.S. Fish and Wildlife Service.

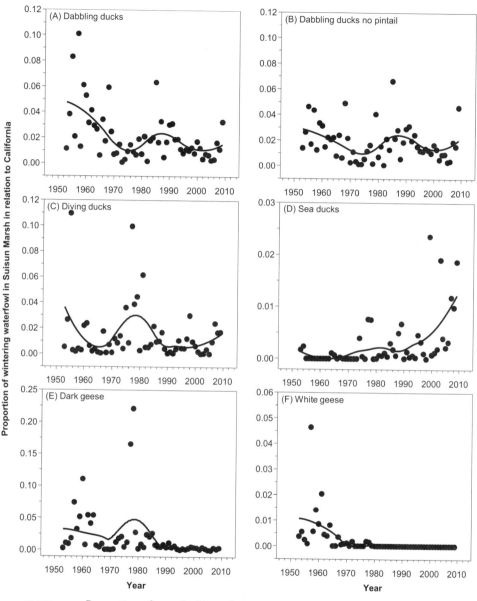

FIGURE 5.3. Proportion of waterfowl in each taxonomic group wintering in Suisun Marsh in relation to the entire state of California, 1953–2009. Waterfowl are categorized as (A) dabbling ducks, (B) dabbling ducks not including pintail, (C) diving ducks, (D) sea ducks, (E) dark geese, and (F) white geese. See table 5.1 for species within each taxonomic group. Solid line illustrates the population trend and was estimated with a LOESS function. Midwinter waterfowl data were collected by the California Department of Fish and Wildlife and U.S. Fish and Wildlife Service. Suisun Marsh was not surveyed as part of the midwinter waterfowl survey in 2010 and 2011.

TABLE 5.2. Percent species composition of waterfowl in Suisun Marsh and California.

Based on (A) midwinter waterfowl surveys and (B) hunter-harvested birds. Midwinter waterfowl surveys were started in 1953 but were not conducted in Suisun Marsh in 2010 or 2011. Hunter-harvest surveys were started in 1961. Midwinter waterfowl survey data are from U.S. Fish and Wildlife Service and California Department of Fish and Game; hunter harvest data are from U.S. Fish and Wildlife Service's Parts Collection Survey.

	Suisun Marsh						
Species[a]	1953–59	1960–69	1970–79	1980–89	1990–99	2000–09	2010–11
(A) MIDWINTER WATERFOWL POPULATION INDEX							
Northern pintail	54.3%	56.6%	48.8%	43.7%	22.4%	19.6%	na
American wigeon	9.7%	7.8%	3.9%	5.5%	10.8%	12.6%	na
Mallard	7.1%	4.9%	7.0%	10.5%	13.1%	13.0%	na
Northern shoveler	5.4%	11.7%	13.9%	21.4%	26.3%	27.8%	na
Greater white-fronted goose	4.2%	4.2%	8.6%	1.9%	0.4%	0.3%	na
American green-winged teal	2.6%	5.9%	1.8%	7.8%	15.6%	12.7%	na
Snow and Ross's goose	1.5%	2.0%	0.4%	0.0%	0.0%	0.0%	na
Cackling goose	1.3%	3.3%	4.6%	0.0%	0.0%	0.0%	na
Gadwall	0.8%	1.1%	0.2%	0.6%	2.7%	3.3%	na
Ruddy duck	0.7%	0.4%	1.1%	2.4%	1.6%	2.6%	na
Canvasback	0.6%	1.2%	7.7%	1.4%	2.3%	0.8%	na
Canada goose	0.3%	0.4%	1.0%	1.4%	1.6%	2.4%	na
Scaup	0.2%	0.0%	0.8%	2.3%	1.1%	2.6%	na
Redhead	0.1%	0.0%	0.0%	0.0%	0.3%	0.0%	na
Cinnamon teal	0.0%	0.1%	0.1%	0.0%	0.0%	0.1%	na
Tundra swan	0.0%	0.1%	0.0%	0.0%	0.5%	0.5%	na
Ring-necked duck	0.0%	0.0%	0.0%	0.2%	0.1%	0.2%	na
Common goldeneye	0.0%	0.0%	0.0%	0.0%	0.5%	0.2%	na
Bufflehead	0.0%	0.0%	0.0%	0.3%	0.2%	1.2%	na
Black brant	0.0%	0.0%	0.0%	0.0%	0.0%	0.0%	na
Eider	0.0%	0.0%	0.0%	0.0%	0.0%	0.0%	na
Common merganser	0.0%	0.0%	0.0%	0.0%	0.0%	0.0%	na
Surf scoter	0.0%	0.0%	0.1%	0.2%	0.1%	0.0%	na
Wood duck	0.0%	0.0%	0.0%	0.0%	0.0%	0.0%	na
(B) HUNTER HARVEST							
Northern pintail	na	46.6%	50.7%	28.3%	13.1%	8.6%	9.3%
American wigeon	na	13.4%	12.2%	13.0%	16.0%	14.4%	13.5%
Northern shoveler	na	11.4%	10.6%	11.4%	14.6%	12.9%	19.9%
Mallard	na	10.6%	8.3%	22.0%	23.7%	25.4%	27.7%
American green-winged teal	na	9.6%	11.1%	14.1%	19.0%	19.7%	17.6%
Gadwall	na	1.4%	1.1%	3.5%	5.2%	6.6%	3.7%
Greater white-fronted goose	na	1.1%	0.6%	0.4%	0.2%	0.9%	1.4%
Canada goose	na	1.1%	0.8%	0.3%	0.5%	2.1%	0.7%
Ruddy duck	na	1.0%	0.8%	0.6%	0.1%	0.4%	1.0%

1953–59	1960–69	1970–79	1980–89	1990–99	2000–09	2010–11
41.5%	39.1%	52.2%	35.7%	25.5%	26.5%	27.6%
14.8%	15.8%	10.6%	11.4%	10.9%	11.5%	7.9%
13.2%	8.5%	8.6%	11.5%	11.2%	7.5%	4.5%
3.9%	5.1%	10.1%	8.9%	9.8%	11.5%	13.3%
2.9%	3.2%	1.5%	1.9%	4.2%	6.6%	7.5%
3.5%	5.7%	3.4%	6.5%	9.5%	8.7%	10.5%
7.1%	10.9%	6.7%	11.4%	12.8%	12.1%	14.2%
2.7%	2.9%	1.1%	0.7%	0.9%	0.6%	1.0%
0.4%	1.1%	0.3%	1.3%	2.7%	3.7%	3.7%
0.8%	1.9%	1.3%	2.4%	2.2%	1.9%	2.4%
1.5%	1.1%	1.0%	1.0%	0.9%	1.1%	1.5%
1.0%	0.5%	0.8%	1.0%	0.9%	0.6%	0.3%
1.8%	2.0%	0.9%	2.4%	2.8%	2.5%	2.0%
0.0%	0.0%	0.0%	0.0%	0.0%	0.0%	0.1%
0.0%	0.1%	0.0%	0.1%	0.2%	0.2%	0.1%
0.3%	0.7%	0.8%	1.4%	1.7%	1.6%	1.3%
0.0%	0.0%	0.0%	0.2%	0.7%	1.2%	1.1%
0.0%	0.1%	0.0%	0.0%	0.1%	0.1%	0.0%
0.1%	0.1%	0.1%	0.5%	0.6%	0.6%	0.3%
0.1%	0.0%	0.0%	0.0%	0.1%	0.1%	0.3%
0.0%	0.0%	0.0%	0.0%	0.0%	0.0%	na
0.0%	0.0%	0.1%	0.1%	0.1%	0.1%	0.0%
0.3%	0.5%	0.4%	1.1%	1.5%	0.7%	0.2%
0.0%	0.1%	0.1%	0.0%	0.0%	0.0%	0.0%

na	31.6%	35.9%	22.5%	11.9%	9.1%	14.2%
na	12.8%	11.1%	10.5%	12.2%	13.7%	13.2%
na	9.1%	8.4%	9.5%	11.0%	12.2%	12.9%
na	18.0%	15.5%	23.1%	26.1%	23.5%	19.4%
na	15.9%	17.7%	19.3%	23.9%	24.4%	23.0%
na	2.7%	2.4%	4.7%	7.0%	7.8%	7.3%
na	3.2%	2.2%	1.2%	1.5%	3.6%	4.0%
na	4.2%	3.8%	3.5%	2.4%	3.3%	4.0%
na	1.2%	1.1%	0.7%	0.4%	0.2%	0.7%

(continued)

TABLE 5.2. *(continued)*

Species[a]	Suisun Marsh						
	1953–59	1960–69	1970–79	1980–89	1990–99	2000–09	2010–11
Cinnamon teal	na	0.9%	0.8%	1.4%	1.5%	2.0%	1.0%
Scaup	na	0.8%	0.5%	1.2%	2.1%	2.1%	0.6%
Canvasback	na	0.7%	1.2%	1.8%	1.1%	0.7%	0.0%
Ring-necked duck	na	0.4%	0.3%	0.5%	0.4%	0.3%	0.6%
Bufflehead	na	0.3%	0.1%	0.4%	0.8%	0.9%	0.8%
Common goldeneye	na	0.2%	0.1%	0.4%	0.8%	1.4%	1.0%
Snow and Ross' goose	na	0.2%	0.2%	0.3%	0.0%	0.3%	0.0%
Redhead	na	0.2%	0.1%	0.2%	0.4%	0.2%	0.0%
Wood duck	na	0.1%	0.1%	0.2%	0.4%	0.4%	0.8%
Black brant	na	0.0%	0.0%	0.0%	0.0%	0.3%	0.0%
Common merganser	na	0.0%	0.0%	0.0%	0.0%	0.0%	0.0%
Surf scoter	na	0.0%	0.0%	0.0%	0.0%	0.0%	0.0%

[a] Some species are not differentiated during midwinter waterfowl surveys, and we have listed the dominant species only. However, canada goose includes lesser, Taverner, dusky, and western canada goose; cackling goose includes Aleutian and small cackling goose; American wigeon may include some Eurasian wigeon; scaup includes both lesser scaup and greater scaup; surf scoter may include some black scoter and white-winged scoter; common goldeneye may include some Barrow's goldeneye; common merganser may include some hooded merganser and red-breasted merganser; and eider includes *Somateria* spp.

the Central Valley. For example, wetlands in the Central Valley have increased as a result of the 1990 implementation of the CVJV (Fleskes et al. 2005b), increased land area used in rice (*Oryza sativa*) production (Fleskes et al. 2005a), and the widespread practice of flooding, rather than burning, rice straw residues for decomposition due to burning restrictions enacted in 1991 (California Rice Straw Burning Reduction Act of 1991).

Most notable is the decline of pintail (figure 5.4A), which historically outnumbered all other species of waterfowl wintering in Suisun Marsh (table 5.2). During the 1950s, an average of 235,800 pintail wintered in the Marsh (table 5.1), accounting for 54% of all waterfowl (table 5.2). In contrast, by the 2000s, only 14,000 pintail, or 20% of waterfowl, wintered in Suisun Marsh (tables 5.1 and 5.2). This dramatic decline reflects, in part, broader declines in pintail at the continental scale (USFWS 2011). In fact, the pintail is the only major dabbling duck species in North America, besides the American black duck (*Anas rubripes*), whose population has remained depressed after the 1986 implementation of the North American Waterfowl Management Plan, which was established in response to record-low waterfowl populations continent wide. Yet, even after accounting for this overall decline, proportionately fewer pintail winter in Suisun Marsh today in relation to the statewide population than historically.

	California					
1953–59	1960–69	1970–79	1980–89	1990–99	2000–09	2010–11
na	3.2%	2.9%	3.3%	3.0%	3.4%	2.8%
na	1.9%	1.4%	1.6%	1.2%	1.4%	0.8%
na	0.7%	1.2%	1.1%	1.1%	0.8%	1.0%
na	0.7%	0.7%	1.4%	1.8%	1.7%	2.2%
na	0.9%	0.4%	0.7%	0.8%	0.8%	0.5%
na	0.4%	0.2%	0.3%	0.3%	0.4%	0.3%
na	3.9%	4.3%	4.0%	4.1%	4.0%	4.1%
na	0.4%	0.5%	0.7%	0.7%	0.6%	0.4%
na	1.1%	1.2%	1.9%	2.3%	2.4%	2.0%
na	0.5%	0.2%	0.0%	0.3%	0.1%	0.0%
na	0.1%	0.1%	0.1%	0.1%	0.1%	0.1%
na	0.1%	0.1%	0.1%	0.0%	0.0%	0.3%

Approximately 13% of California's pintail used the Marsh in the 1950s, but that percentage has steadily declined to only 1% in the 2000s (table 5.1). Therefore, the dramatic decline in the pintail population index in the Marsh reflects both the decline in the continent-wide population and fewer pintail wintering in the Marsh than in other parts of the state. This decline of pintail has led many land managers to prioritize management for other waterfowl species and change wetland management in the Marsh from the shallow open-water habitats preferred by pintail to the more vegetated marsh habitats preferred by other dabbling ducks, such as mallard (S. Chappell, personal communication).

In contrast to dabbling ducks and geese, diving duck abundance has remained relatively stable (figure 5.1C) and sea ducks have increased within the Marsh (figure 5.1D). The increase in sea ducks is attributable to bufflehead, which have increased in the Marsh proportionately faster than their overall increase within the state. Nonetheless, overall population sizes of sea ducks and diving ducks are much smaller in the Marsh than current populations of dabbling ducks (table 5.1).

Presently, 90% of waterfowl wintering in Suisun Marsh are dabbling ducks, followed by diving ducks (5%), geese (2%), and sea ducks (1%; table 5.1). Furthermore, the majority of diving ducks and sea ducks occur at Joice Island, a deep-water wetland unit adjacent to Montezuma Slough and Suisun Slough, and on Mallard

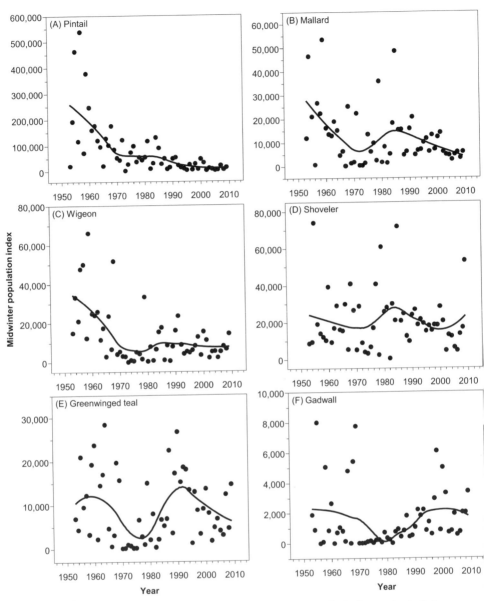

FIGURE 5.4. Midwinter population size indices for dabbling ducks by species in Suisun Marsh, 1953–2009. Dabbling ducks are (A) northern pintail, (B) mallard, (C) American wigeon, (D) northern shoveler, (E) American green-winged teal, and (F) gadwall. Solid line illustrates the population trend and was estimated with a LOESS function. Midwinter waterfowl data were collected by the California Department of Fish and Wildlife and U.S. Fish and Wildlife Service. Suisun Marsh was not surveyed as part of the midwinter waterfowl survey in 2010 and 2011.

Reservoir, located south of Suisun Bay just north of Concord. Therefore, present management in the Marsh remains focused on managing wetlands for dabbling ducks.

It is unlikely that increases in diving ducks or sea ducks could compensate for any continued decline of dabbling ducks. Diving ducks and sea ducks require distinctly different habitat than dabbling ducks, preferring deeper water and having a diet that favors invertebrates (including bivalves) rather than plants. Interestingly, coots also have declined in abundance within Suisun Marsh (table 5.1). Coots are a species of rail but, like dabbling ducks, predominantly use freshwater and brackish wetlands. While California populations of coot have more than tripled in the past 25 years, the wintering population of coot within the Marsh has not changed over the same period. The decline of coots along with the decline of dabbling ducks in the Marsh indicates larger landscape changes.

Waterfowl Habitat Use

The vast majority of waterfowl that winter in Suisun Marsh are dabbling ducks (table 5.1), and they primarily use managed wetland habitats provided by duck hunting clubs and state wildlife areas. Casazza et al. (2012) radiomarked female pintail and tracked them each winter from 1990 to 1993. They found that the habitats used varied between day, when ducks typically roost and rest, and night, when ducks primarily forage. Pintail selected undisturbed sanctuaries during the day and preferred managed wetland habitats containing a relatively high density of brass buttons (*Cotula coronopifolia*) at night. There also is evidence for limited movements of ducks between Suisun Marsh and the nearby Delta region (Casazza 1995; Miller et al. 2009).

We reanalyzed data obtained with radiomarked pintail specifically to examine habitat selection by ducks in Suisun Marsh. To do so, we compared the spatial patterns of habitat use by ducks to the availability of those habitats at two spatial scales.[7] We estimated habitat use as the proportion of locations of radiomarked pintail found within each habitat type. We found that pintail strongly selected managed wetland habitats at both small and large scales of analysis, with use of managed wetland habitat much greater than its availability; pintail avoided tidal marshes, bays and sloughs, and other habitats (see map 10 in color insert). These

---

7. We estimated habitat availability at two spatial scales, using the hierarchical ordering process suggested by Johnson (1980) for habitat selection studies. For first-order habitat selection, we considered all the area contained in the Suisun Marsh Basin boundary to be potentially available to ducks. For third-order habitat selection, we considered all the area contained in the pintail population's home range, as determined by minimum convex polygon using all 7,825 telemetry locations from the 215 radiomarked pintails. We categorized habitats as tidal marsh, managed wetlands, bays and waterways, and other habitat (which included uplands and grazed lands) using the Bay Area EcoAtlas habitat categories (San Francisco Estuary Institute 1998).

findings are consistent with dabbling duck habitat associations for other areas with tidal marsh habitat. For example, along the coast of South Carolina, the occurrence of dabbling ducks in managed coastal freshwater impoundments was significantly greater than expected by chance, while the occurrence of dabbling ducks in unmanaged tidal wetland habitats was significantly less than expected (Gordon et al. 1998).

Waterfowl Diet

Similar to most managed wetland areas within the Central Valley, Suisun Marsh wetlands are managed primarily for waterfowl food production, especially food for dabbling ducks. However, brackish water in the Marsh makes it difficult to manage for plant species that tend to be highly productive and provide abundant energy-rich seeds sought by waterfowl in the Central Valley. Early studies of waterfowl food plants revealed that most ducks in Suisun Marsh fed principally on seeds of several moist-soil plant species. In particular, two studies discovered that ducks used alkali bulrush (also called "salt marsh bulrush"; *Bolboschoenus maritimus*) and the alien species brass buttons more than other foods present in the Marsh (George et al. 1965; Mall 1969).

On the basis of these studies, biologists developed plans for managing salinity levels that would enable growth of these key waterfowl food plants in wetlands (Rollins 1973, 1981; Miller et al. 1975). The plans were based largely on findings that the duration of soil submergence and soil salinity are the two primary factors that most influence vegetation growth (Mall 1969; Rollins 1973). Because of the seasonally brackish nature of Suisun Marsh, careful management was required to prevent high soil salinities. Rollins (1981) found that circulation of water and multiple leach cycles in spring reduced salinity in plants' root zones and, coupled with specific flooding durations, encouraged growth of alkali bulrush, fat hen (*Atriplex prostrata*), and brass buttons. Accordingly, salinity standards for the Marsh were set to ensure desired soil salinity conditions for waterfowl food plants (Suisun Ecological Workgroup 2001).

More recent studies (Burns 2003; Burns et al. 2003) found that ducks consumed more than 30 species of plant seeds, with seeds of only 10 species accounting for more than 90% of the diet. The most important food plants for ducks included alkali bulrush, watergrass (*Echinochloa crus-galli*), and western sea purslane (*Sesuvium verrucosum*), although their importance varied among duck species (percent aggregate dry mass: 37% sea purslane, 24% alkali bulrush, 21% watergrass, 6% fat hen, 4% swamp timothy [*Crypsis schoenoides*], 3% pickleweed [*Salicornia pacifica*], 2% brass buttons, and 1% each for smartweeds [*Persicaria* spp.], rabbit's foot grass [*Polypogon monspeliensis*], and docks [*Rumex crispus, R. occidentalis*]). Ducks also consumed more than 20 invertebrate taxa, but they composed a very small proportion of the total diet (less than 5%). Overall, these diet studies confirmed the importance of moist-soil seed plants for dabbling ducks in Suisun Marsh.

OVERVIEW: BREEDING WATERFOWL

Waterfowl are highly migratory, and most waterfowl that winter within the United States originate from breeding grounds in Alaska and Canada. California is unique among wintering areas for North American waterfowl, in that a large percentage of mallard, gadwall, and cinnamon teal are produced locally on breeding grounds within the state. In fact, over 60% of mallard harvested in California originate in California; the remaining birds originate from Oregon, Washington, British Columbia, Alberta, Alaska, and the Yukon Territory (Munro and Kimball 1982; Ackerman et al. 2010).

The breeding population index of dabbling ducks in California has averaged 562,000 ducks, of which 364,000 are mallard, since the annual breeding population survey[8] was established in 1992 (figure 5.5). In addition to mallard (62%), these ducks include gadwall (15%), cinnamon teal (8%), shoveler (6%), wood duck (1%), and pintail (1%). On average, 5% of California's surveyed mallard and 7% of gadwall breed in Suisun Marsh. The larger areas in the Sacramento Valley (36%), northeastern California (25%), and San Joaquin grasslands (16%) support the majority of mallard (other areas are East Valley, 6%; San Joaquin desert, 4%; Napa marshes, 4%; San Francisco Bay and Sacramento–San Joaquin Delta, 2%; and West Valley, 2%).[9]

The *proportions* of mallard and gadwall breeding in Suisun Marsh in relation to their statewide breeding populations have declined significantly since 1992 (figure 5.5). Nevertheless, breeding *densities* of ducks in Suisun Marsh are still among the highest in the state. The Marsh is considered a hot spot for high nesting densities, even compared with more traditional nesting areas in North America such as the Prairie Pothole Region in the central United States and Canada. In addition to occurring in high nesting densities, ducks in Suisun Marsh also tend to initiate nesting attempts earlier in the breeding season than elsewhere in North America, thus yielding more opportunities for hens to produce a successful clutch during the summer (McLandress et al. 1996; Krapu et al. 2002).

One of the world's longest-running studies on nesting ecology of ducks has been conducted annually in Suisun Marsh since 1985 at the Grizzly Island Wildlife Area (Ackerman et al. 2009). The majority of breeding ducks are mallard (75%), gadwall (17%), pintail (5%), cinnamon teal (2%), and shoveler (1%). However,

---

8. Breeding-population survey data were collected by the California Department of Fish and Game and were kindly provided by Dan Yparraguirre and Shaun Oldenburger.

9. Mean breeding population size index, 1992–2011.

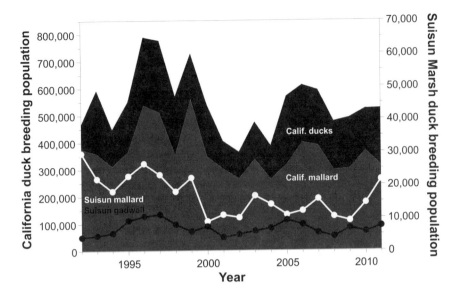

FIGURE 5.5. Breeding population index for dabbling ducks in California (left axis) and Suisun Marsh (right axis), 1992–2011. The filled black polygon represents the population size estimate for dabbling ducks in California, and the filled gray polygon represents the population size estimate for mallards in California (left axis). The thin white line represents the population size estimate for mallard in the Marsh, and the thin black line represents the population size estimate for gadwall in the Marsh (right axis). Species included in the dabbling duck category were mallard, gadwall, pintail, cinnamon teal, and shoveler. Breeding population surveys were conducted by the California Department of Fish and Wildlife.

in recent years the proportion of nesting mallard has been much lower (60%), while the proportion of gadwall (33%) has increased. This upland habitat used for nesting by ducks also is used for nesting by northern harriers, short-eared owls, American bittern, ring-necked pheasant, and numerous songbirds. The study documented a dramatic increase in dabbling duck nesting density, up to 16 nests/ha, followed by a precipitous decline (Ackerman et al. 2009). The decline in duck productivity may be associated with upland vegetation quality. In the 1980s and early 1990s, the California Department of Fish and Game actively managed vegetation in several of the fields by planting various mixes of tall wheat grass (*Elymus ponticus*), vetch (*Vicia* spp.), alfalfa (*Medicago sativa*), perennial rye grass (*Festuca* spp.), and wheat (*Elymus* spp.) that produced high-quality, dense nesting cover for ducks. Ducks seemed to respond positively to this active upland-vegetation management, and nesting densities increased until they peaked in

1997. During February 1998, a series of strong El Niño storms flooded the Marsh (including Grizzly Island Wildlife Area) as a result of levee breaches and overtopping. Brackish water subsequently intruded and much of the vegetation used as nesting cover was flooded, and in the following years the habitat deteriorated. Concurrently, nesting densities declined after 1997.

Nest survival is highly variable in Suisun Marsh (McLandress et al. 1996; Ackerman 2002a, 2002b; Ackerman et al. 2003, 2004), as it is in most duck nesting areas, but has shown a general decline since 1985 (Ackerman et al. 2009). Still, in most years, nest survival has been above the estimated 15% to 20% nest success required to maintain a stable mallard population (Cowardin et al. 1985) and is among the highest reported for ducks in North America (McLandress et al. 1996). The reason for the decline of nest density and nest survival most likely involves interactions among predators, alternative prey resources, and habitat quality. Since 2008, vegetation at Grizzly Island Wildlife Area has been undergoing large-scale restoration of upland plant communities, and water delivery infrastructure has been improved in an attempt to improve waterfowl breeding opportunities (Ackerman et al. 2009). Although it may take several years to assess breeding ducks' responses to these changes, a rapid increase in nest densities and improved nest survival have been documented.

After breeding, waterfowl lose and regrow their flight feathers (molt) over about three weeks in late summer, during which time they are flightless. This is a critical period because the ducks are more susceptible to predation, physiological stress, and disease. Most mallard that breed in the Marsh do not molt there but instead migrate northward to the Sacramento Valley and Klamath Basin (Yarris et al. 1994). There, large wetlands are traditionally flooded during summer and are dominated by tall emergent vegetation, such as bulrush (*Schoenoplectus* spp.) and cattail (*Typha* spp.), which can provide important cover, protection, and a reliable food source during this vulnerable time. The molt migration from Suisun Marsh suggests that predictably available molting habitats within the Marsh are limited. This further suggests that efforts to improve and expand molting habitat could be successful management actions to sustain waterfowl populations (Fleskes et al. 2010).

The widespread contribution of local reproduction of ducks in Suisun Marsh is illustrated by band recoveries from hunter-killed ducks. Dabbling ducks, primarily mallard, have been banded during summer (May–September), and more than 9,000 of these ducks have been recovered throughout North America since 1932 (see map 11 in color insert).[10] These banding data illustrate the extensive

10. Ducks were banded within the Suisun Marsh by California Waterfowl and California Department of Fish and Game, and band recoveries were thereafter managed by the U.S. Geological Survey's Bird Banding Lab.

connectedness of Suisun Marsh ducks to the Pacific Flyway and throughout the continent's dabbling duck populations.

DATA GAPS

There are numerous data gaps for wintering and breeding waterfowl in Suisun Marsh, and we highlight five major areas of needed research below.

If a portion of Suisun Marsh is restored to tidal marsh, will there be enough food to maintain the present size of waterfowl populations? The CVJV estimates that converting 2,025 ha (5,000 acres) of managed wetlands to tidal marsh in Suisun Marsh could reduce food resources below those necessary to support the wintering population of dabbling ducks by mid-January (CVJV 2006). However, because detailed information about food values of brackish seasonal and tidal wetlands in the Marsh is lacking, this estimated model projection should be refined when more precise estimates of seed production, abundance of vegetation and invertebrate food resources, available energy of seed and other foods, and resource availability (e.g., water depth) to waterfowl become available.

Will ducks use tidal marsh habitats to the same extent that they use managed wetlands? When the Marsh was relatively less altered in the late 1800s, wintering waterfowl abundance was substantially greater than at present. However, under the highly altered landscape of today, most dabbling ducks do not use tidal marsh habitats (see map 10 in color insert). Whether or not ducks would use tidal areas within Suisun Marsh if their extent increased markedly is questionable, because overall salinity levels have already increased as a result of water diversions, and waterfowl food production within tidal marshes is considerably lower than in diked, brackish wetlands.

Will Suisun Marsh continue to provide habitat for some of the highest nest densities and reproductive success of breeding ducks in California? Suisun Marsh provides some of the highest nesting densities of ducks in California and contributes a relatively large proportion of ducks to the state in relation to its size. How the future habitat mosaic will affect ducks' nest densities and reproductive success if waterfowl habitat configurations in the Marsh are altered is unknown.

Would an increase in tidal marsh and overall salinity levels in wetlands reduce ducks' reproductive success? Considerable evidence indicates that duckling growth and survival can be affected by water salinity levels because salt glands of ducklings are poorly developed at hatching. For example, ducklings subjected to water salinity as low as 4 millisiemens per cm (mS/cm) have impaired growth rates, whereas 20 mS/cm is lethal to mallard ducklings (Mitcham and Wobeser 1988a, 1988b). Survival rates of ducklings and shorebird chicks are not known for Suisun Marsh, but these data will be critical for evaluating future habitat restoration plans.

What are the societal impacts of having fewer waterfowl in Suisun Marsh? The fact that Suisun Marsh is the largest contiguous brackish marsh in the western United States owes much to the conservationist legacy of duck hunting. Suisun Marsh wetlands have been preserved through the duck hunting heritage at considerable expense, with 75% of wetlands in Suisun Marsh under private ownership (Gill and Buckman 1974). The societal impacts of returning portions of Suisun Marsh to tidal marsh are important, although some conversion is inevitable with sea-level rise and the increasing pressure on the Delta's levee system (see chapter 3). We urge a collaborative approach to Suisun Marsh restoration and maintaining the strong stewardship that duck hunters have provided to the Marsh since the turn of the 20th century.

THE FUTURE

Carrying Capacity for Waterfowl

Ultimately, the goal of understanding waterfowl population trends and diets is not only to improve habitat management programs, but also to determine the capacity of Suisun Marsh to support current (or desired) populations of waterfowl. To establish wintering habitat objectives, the CVJV (one of the original six priority joint ventures formed under the North American Waterfowl Management Plan) uses a bioenergetic model that calculates the energy requirements of a waterfowl population. The model is useful for assessing current habitat conditions for wintering waterfowl, and it also can be used to predict how changes in policy, land use, or habitat programs might affect waterfowl (CVJV 2006).

The CVJV waterfowl population goal is a population size comparable to those of the mid-1970s under the directive of the North American Waterfowl Management Plan. Suisun Marsh supported up to 300,000 waterfowl (mainly dabbling ducks) at peak abundance by late December in the 1970s, with numbers tapering to 50,000 by mid-February. Estimating the energy supply (food abundance and caloric value) for waterfowl in Suisun Marsh is much more difficult, because systematic studies have not been conducted. However, several studies have suggested that plants in brackish or tidal saline habitats have lower seed abundance, produce seeds of lower metabolizable energy, and result in waterfowl in poorer body condition (Mall 1969; Atwater et al. 1979; Tietje and Teer 1988, 1996; Ballard et al. 2004). Therefore, the CVJV developed planning models for Suisun Marsh under several scenarios. One model assumed that moist-soil seed production in Suisun Marsh was equivalent to seed production in freshwater marshes (566 lbs/acre). An alternative model used an estimate of seed production that was half that of freshwater marshes (283 lbs/acre). Under both scenarios, the CVJV then incorporated potential effects of the conversion of 2,025 ha of managed wetlands into tidal marsh (CVJV 2006).

The bioenergetic model's outcomes under these different scenarios are profoundly different. Under the assumption that brackish wetlands in Suisun Marsh have similar food production as freshwater wetlands, food supply is more than adequate to meet the foraging needs of desired waterfowl populations; this remains true even if, for example, 2,025 ha are restored to tidal marsh. However, if food production in tidal marsh is half that of managed freshwater wetlands, then the food supply could be exhausted by early spring. Furthermore, the impact of converting 2,025 ha of managed wetlands to tidal marsh under this scenario could be large, with food demand exceeding available supply by mid-January. Thus, the impact of tidal marsh restoration on waterfowl from a bioenergetics perspective may be of little or of great consequence, depending largely on baseline levels of seed production and quality in both managed and tidal marshes. Such results suggest that alternative management scenarios need to be considered, such as concentrating efforts (e.g., levee maintenance) in areas most defensible from sea-level rise, land subsidence, and other factors (see chapter 9).

The Past and Future of Waterfowl

If we look far enough into the future, sea-level rise and levee failures will likely flood some portion of the currently diked wetland landscapes in Suisun Marsh (Knowles 2010; Stralberg et al. 2011). In the near term (e.g., this century), reinforcing levees, raising levee crown heights, using large water pumps, and managing soil salinity levels with intensive leaching strategies may be able to overcome some flooding events and salinity intrusion, and it may be possible to retain many diked wetlands.

If the immediate future for Suisun Marsh includes restoring a portion of it to a tidal system to prepare for future conditions, it is important to understand how this will affect the carrying capacity of Suisun Marsh and the waterfowl populations that depend on these resources. Yet this necessary information is incomplete and uncertain. We know that nearly 370,000 waterfowl wintered in Suisun Marsh in the past, and we know that the plant community composition was undoubtedly different. Atwater et al. (1979: 369), for example, described the tidal marsh plants of Carquinez Strait and Suisun Bay as "an intricate, mutable transition between the salt marshes of San Francisco Bay and the freshwater marshes of the

Delta." Plant species that prevailed in the transitional tidal zone of Suisun Marsh that may have provided food value to waterfowl included pickleweed, various bulrushes, brass buttons, rabbit's foot grass, and lamb's quarters (*Chenopodium album;* Atwater et al. 1979). Invertebrates and bivalves are important to diving ducks and sea ducks, but dabbling ducks rely on invertebrates primarily during late winter and early spring when they require protein for migration and breeding. Therefore, with the exception of alkali bulrush and potentially sago pondweed (*Stuckenia pectinata;* Moffitt 1938), it is unclear what food plants might have produced sufficient seeds for the large numbers of waterfowl in Suisun Marsh prior to the 1850s.

Moreover, it is unclear how the existing food base would change as a result of tidal marsh restoration efforts. For example, higher salinity levels during drought periods in Suisun Marsh substantially reduced the growth and seed production of alkali bulrush (Mall 1969; Atwater et al. 1979). Studies in Gulf Coast marshes have shown that foods consumed in relatively saline habitats by pintail contained less protein and fat than foods consumed in inland freshwater wetlands and flooded rice fields, and that ducks in saline environments had lower body condition (Ballard et al. 2004). Presently, body condition of pintail and mallard are slightly better in Suisun Marsh than in the Delta, which suggests that saline conditions are not currently affecting body condition of adult ducks within the Marsh (Miller et al. 2009).

That waterfowl populations were historically much larger in Suisun Marsh may not be at odds with the fact that more saline conditions likely favored plant species with lower seed yields. For example, food resources may have been limited when considered on a *per unit area* basis, but this low productivity may have been more than compensated for by the historically extensive wetlands that occurred in the Suisun Marsh, Delta, and Central Valley. The expansiveness of California's wetland areas is long past, with 91% of wetlands in California lost to agriculture and urban development (Dennis and Marcus 1984; Dahl 1990). To support the remaining waterfowl populations, we may not be able to simply turn back the clock with a desired habitat end-state mimicking the tidal marshes before the Gold Rush. California has been so highly modified that active, and perhaps even intense, wetland management may be required to maintain the diversity and abundance of waterfowl and other wildlife.

A similar conundrum exists just down river in San Francisco Bay, where former salt evaporation ponds provide habitat for most of the migrating, wintering, and breeding shorebirds, seabirds, and waterfowl that use the Bay. The restoration of these salt ponds to tidal marsh is being balanced with a mosaic of intensively managed wetlands for waterbird habitat (see www.southbayrestoration.org/). In both San Francisco Bay and Suisun Marsh, the question is the same: *How do we provide for the historical diversity and abundance of animals that existed in the past within a tidal marsh system that has been drastically reduced in size? The* answer may simply be that we cannot. Thus, having a marsh mosaic that includes managed wetlands with their higher productivity and tidal marshes with their more natural state may be the only way to both restore tidal marsh habitat and maintain the diversity and abundance of animals that once used these habitats. The intellectual and logistical challenge in Suisun Marsh is to find an array of solutions that balance the needs of waterfowl with the myriad other ecological services provided by the Marsh.

SUMMARY

Five million waterfowl, representing 68% of waterfowl in the Pacific Flyway, winter within California. Suisun Marsh hosts over 60,000 of these waterfowl and plays an important role in relation to its size. Current waterfowl abundance is below the population objective for 300,000 ducks wintering in Suisun Basin, and well below the ~370,000 waterfowl that wintered there during the 1950s. Long-term population trends for dabbling ducks and geese are declining in the Marsh, but diving ducks and swans are stable, and sea ducks are slightly increasing. These same trends occur after accounting for California-wide population trends, indicating that proportionately fewer dabbling ducks and geese are choosing to winter in Suisun Marsh than historically. Most notable is the decline of pintail. More than 235,000 pintail accounted for 54% of all waterfowl using the Marsh in the 1950s, but their numbers had declined to only 14,000, or 20% of all waterfowl, by the 2000s. The decline of pintail led many land managers to change wetland management from shallow open-water habitats preferred by pintail to more vegetated habitats. The increase in sea ducks in Suisun Marsh is attributable to bufflehead, which have increased in the Marsh disproportionately faster than their increase within the state. Dabbling ducks presently—as they did historically—account for 90% of wintering waterfowl in Suisun Marsh, followed by diving ducks (5%), geese (2%), sea ducks (1%), and swans (1%). Dabbling ducks strongly select managed wetland habitats and avoid tidal marshes, bays, and sloughs, feeding mainly on seeds of alkali bulrush, watergrass, and sea purslane. Suisun Marsh also supports among the highest densities of breeding ducks in California. Approximately 5% of California's 364,000 surveyed mallard and 7% of California's 90,000 gadwall breed in the relatively small area of the Marsh. Ducks banded in the Marsh are recovered throughout the Pacific Flyway and North America, which illustrates the Marsh's connectedness to the continent's waterfowl populations.

With more than 90% of California's wetlands lost, Suisun Marsh represents one of the last remaining contiguous wetland habitats in the state. Waterfowl have long been present in the Marsh in considerably larger abundance. Maintaining the present, if not the historical, diversity and abundance of waterfowl in the Marsh likely will depend on active wetland management for higher-yielding seed plants that can increase the carrying capacity of the few remaining wetlands.

REFERENCES

Ackerman, J. T. 2002a. From individuals to populations: the direct and indirect effects of predation on waterfowl nest success. PhD dissertation, University of California, Davis.

Ackerman, J. T. 2002b. Of mice and mallards: positive indirect effects of coexisting prey on waterfowl nest success. Oikos 99: 469–480.Ackerman, J. T., A. L. Blackmer, and J. M. Eadie. 2004. Is predation on waterfowl nests density dependent? Tests at three spatial scales. Oikos 107: 128–140.

Ackerman, J. T., J. M. Eadie, D.. Loughman, G. S. Yarris, and M. R. McLandress. 2003. The influence of partial clutch depredation on duckling production. Journal of Wildlife Management 67: 576–587.

Ackerman, J. T., C. A. Eagles-Smith, G. Herring, and M. P. Herzog. 2010. Feather findings: stable isotope signatures reveal the origins of California mallards. California Waterfowl (fall/winter): 18–19.

Ackerman, J. T., J. Kwolek, R. Eddings, D. Loughman, and J. Messerli. 2009. Evaluating upland habitat management at the Grizzly Island Wildlife Area: effects on dabbling duck nest density and nest success. Administrative report. U.S. Geological Survey, Western Ecological Research Center, Davis, California, and California Waterfowl Association, Sacramento.

Ackerman, J. T., J. Y. Takekawa, D. L. Orthmeyer, J. P. Fleskes, J. L. Yee, and K. L. Kruse. 2006. Spatial use by wintering greater white-fronted geese relative to a decade of habitat change in California's Central Valley. Journal of Wildlife Management 70: 965–976.

Arnold, A. 1996. Suisun Marsh History: Hunting and Saving a Wetland. Marina, California: Monterey Pacific.

Atwater, B. F., S. G. Conard, J. N. Dowden, C. W. Hedel, R. L. MacDonald, and W. Savage. 1979. History, landforms, and vegetation of the Estuary's tidal marshes. Pages 347–385 in San Francisco Bay: The Urbanized Estuary (T. J. Conomos, A. E. Leviton, and M. Berson, eds.). San Francisco: AAAS Pacific Division.

Ballard, B. M., J. E. Thompson, J. J. Petrie, M. Chekett, and D. G. Hewitt. 2004. Diet and nutrition of northern pintails wintering along the southern coast of Texas. Journal of Wildlife Management 68: 371–382.

Burns, E. G. 2003. An analysis of food habitat of green-winged teal, northern pintail, and mallards wintering in the Suisun Marsh to develop guidelines for food plant management. MS thesis, University of California, Davis.

Burns, E. G., J. M. Eadie, and M. R. Miller. 2003. Food habits of green-winged teal, northern pintails and mallards wintering in the Suisun Marsh. Final Report. California Department of Water Resources, Sacramento, and U.S. Bureau of Reclamation, Washington, D.C.

Casazza, M. L. 1995. Habitat use and movements of Northern Pintails wintering in the Suisun Marsh, California. MS thesis, California State University, Sacramento.

Casazza, M.L., P.S. Coates, M.R. Miller, C.T. Overton, and D.Y. Yparraguirre. 2012. Hunting influences the diel patterns in habitat selection by northern pintails *Anas acuta*. Wildlife Biology 18: 1–13.

Central Valley Joint Venture. 2006. Central Valley Joint Venture Implementation Plan—Conserving Bird Habitat. U.S. Department of the Interior, Fish and Wildlife Service, Sacramento.

Collins, D.P., C.A. Palmer, and R.E. Trost. 2011. Pacific Flyway Data Book: Waterfowl Harvests and Status, Hunter Participation and Success in the Pacific Flyway in the United States. U.S. Department of the Interior, Fish and Wildlife Service, Portland, Oregon.

Cowardin, L.M., D.S. Gilmer, and C.W. Schaiffer. 1985. Mallard recruitment in the agricultural environment of North Dakota. Wildlife Monographs 92: 1–37.Dahl, T.E. 1990. Wetland losses in the United States 1780s to 1980s. U.S. Department of the Interior, Fish and Wildlife Service, Washington, D.C.

Dennis, N.B., and M.L. Marcus. 1984. Status and trends of California wetlands. Final report prepared for the California Assembly, Resources Subcommittee, Sacramento.

Fleskes, J.P., D.M. Mauser, J.L. Lee, D.S. Blehert, and G.S. Yarris. 2010. Flightless and post-molt survival and movements of female mallards molting in Klamath Basin. Waterbirds 33: 208–220.

Fleskes, J.P., W.M. Perry, K.L. Petrik, R. Spell, and F. Reid. 2005a. Change in area of winter-flooded and dry rice in the northern Central Valley of California determined by satellite imagery. California Fish and Game 91: 207–215.

Fleskes, J.P., J.L. Yee, M.L. Casazza, M.R. Miller, J.Y. Takekawa, and D.L. Orthmeyer. 2005b. Waterfowl distribution, movements and habitat use relative to recent habitat changes in the Central Valley of California: A cooperative project to investigate impacts of the Central Valley Habitat Joint Venture and changing agricultural practices on the ecology of wintering waterfowl. Published Final Report. U.S. Geological Survey Western Ecological Research Center, Dixon Field Station, Dixon, California.

Garone, P. 2011. The Fall and Rise of the Wetlands of California's Great Central Valley. Berkeley: University of California Press.

George, H.A., W. Anderson, and H. McKinnie. 1965. An evaluation of the Suisun Marsh as a waterfowl area. Administrative Report 20. California Department of Fish and Game, Sacramento.

Gill, R., and A.R. Buckman. 1974. The natural resources of Suisun Marsh; their status and future. Coastal Wetland Report 9. California Department of Fish and Game, Sacramento.

Gordon, D.H., B.T. Gray, and R.M. Kaminski. 1998. Dabbling duck–habitat associations during winter in coastal South Carolina. Journal of Wildlife Management 62: 559–580.

Hall, F. 2007. A look back: the Suisun Marsh. California Waterfowl (November–December): 70–73.

Hall, F. 2011. They Came to Shoot: A History of California Duck Clubs and Wetland Conservation. Roseville: California Waterfowl Association.

Johnson, D.H. 1980. The comparison of usage and availability measurements for evaluating resource preference. Ecology 61: 65–71.

Knowles, N. 2010. Potential inundation due to rising sea levels in the San Francisco Bay region. San Francisco Estuary and Watershed Science 8(1): 1–19.

Krapu, G. L., G. A. Sargeant, A. E. H. Perkins. 2002. Does increasing daylength control seasonal changes in clutch sizes of northern pintails (*Anas acuta*)? Auk 119: 498–506.

Mall, R. E. 1969. Soil-water-salt relationships of waterfowl food plants in the Suisun Marsh of California. Wildlife Bulletin No. 1. California Department of Fish and Game, Sacramento.

McLandress, M. R., G. S. Yarris, A. E. H. Perkins, D. P. Connelly, and D. G. Raveling. 1996. Nesting biology of California mallards. Journal of Wildlife Management 60: 94–107.

Miller, A. W., R. S. Miller, H. C. Cohen, and R. F. Schultze. 1975. Suisun Marsh Study, Solano County, California. U.S. Department of Agriculture Soil Conservation Service, Washington, D.C.

Miller, M. R., E. G. Burns, B. E. Wickland, and J. M. Eadie. 2009. Diet and body mass of wintering ducks in adjacent brackish and freshwater habitats. Waterbirds 32: 374–387.

Mitcham, S. A., and G. Wobeser. 1988a. Effects of sodium and magnesium-sulfate in drinking-water on mallard ducklings. Journal of Wildlife Diseases 24: 30–44.

Mitcham, S. A., and G. Wobeser. 1988b. Toxic effects of natural saline waters on mallard ducklings. Journal of Wildlife Diseases 24: 45–50.

Moffitt, J. 1938. Environmental factors affecting waterfowl in the Suisun area, California. Condor 40: 76–84.

Munro, R. E., and C. F. Kimball. 1982. Population ecology of the mallard: VII. Distribution and derivation of the harvest. U.S. Department of Interior, Fish and Wildlife Service Resource Publication No. 147.

Neter, J., M. H. Kutner, C. J. Nachtsheim, and W. Wasserman. 1996. Applied Linear Statistical Models, 4th ed. Chicago: IRWIN.

Olofson, P. R., ed. 2000. Baylands ecosystem species and community profiles: life histories and environmental requirements of key plants, fish and wildlife (P. R. Olofson, ed.). Prepared by the San Francisco Bay Area Wetlands Ecosystem Goals Project. San Francisco Bay Regional Water Quality Control Board, Oakland, California.

Rollins, G. L. 1973. Relationships between soil salinity and the salinity of applied water in the Suisun Marsh of California. California Fish and Game 59: 5–35.

Rollins, G. L. 1981. A Guide to Waterfowl Habitat Management in Suisun Marsh. California Department of Fish and Game, Sacramento.

San Francisco Estuary Institute. 1998. Bay Area EcoAtlas Narrative Documentation, EcoAtlas beta release, version 1.5b. Oakland, California: San Francisco Estuary Institute.

Stoner, E. A. 1934. Summary of a record of duck shooting on the Suisun Marsh. Condor 36: 105–107.

Stoner, E. A. 1937. A record of twenty-five years of wildfowl shooting on the Suisun Marsh, California. Condor 39: 242–248.

Stralberg, D., M. Brennan, J. C. Callaway, J. K. Wood, L. M. Schile, D. Jongsomjit, M. Kelly, V. T. Parker, and S. Crooks. 2011. Evaluating tidal marsh sustainability in the face of sea-level rise: a hybrid modeling approach applied to San Francisco Bay. PLoS ONE 6: e27388.

Suisun Ecological Workgroup. 2001. Final report to the State Water Resources Control Board. California Department of Water Resources, Sacramento.

Swanson, G. A., and J. C. Bartonek. 1970. Bias associated with food analysis in gizzards of blue-winged teal. Journal of Wildlife Management 34: 739–746.

Swiderek, P. K., A. S. Johnson, P. E. Hale, and R. L. Joyner. 1988. Production, management, and waterfowl use of sea purslane, gulf coast muskgrass, and widgeongrass in brackish impoundments. Pages 441–457 *in* Waterfowl in Winter: Proceedings of the 1985 Symposium (M. W. Weller, ed.). Minneapolis: University of Minnesota Press.

Tietje, W. D., and J. G. Teer. 1988. Winter body condition of northern shovelers on freshwater and saline habitats. Pages 353–376 *in* Waterfowl in Winter: Proceedings of the 1985 Symposium (M. W. Weller, ed.). Minneapolis: University of Minnesota Press.

Tietje, W. D., and J. G. Teer. 1996. Winter feeding ecology of northern shovelers on freshwater and saline wetlands in south Texas. Journal of Wildlife Management 60: 843–855.

U.S. Fish and Wildlife Service. 2011. Waterfowl population status, 2011. U.S. Department of the Interior, Washington, D.C.

Yarris, G. S., M. R. McLandress, and A. E. H. Perkins. 1994. Molt migration of postbreeding female mallards from Suisun Marsh, California. Condor 96: 36–45.

GEOSPATIAL DATA SOURCES

CalAtlas. 2012. California Geospatial Clearinghouse. State of California. Available: http://atlas.ca.gov/. Accessed: 2012.

Contra Costa County. 2013. Contra Costa County Mapping Information Center. Available: http://www.ccmap.us/. Accessed: January 2013.

Gesch, D., M. Oimoen, S. Greenlee, C. Nelson, M. Steuck, and D. Tyler. 2002. The National Elevation Dataset. Photogrammetric Engineering and Remote Sensing 68 (1): 5–11.

San Francisco Estuary Institute. 2012. Bay Area EcoAtlas. Available: http://www.sfei.org/ecoatlas/. Accessed: March 20 2012.

Solano County. 2013. Geographic Information Systems Homepage. Solano County Department of Information Technology. Available: http://www.co.solano.ca.us/depts/doit/gis/. Accessed: November 2012.

6

Terrestrial Vertebrates

Alison N. Weber-Stover and Peter B. Moyle

Tidal wetlands comprise roughly 45,000 km^2 globally and are present in isolated pockets or relatively narrow bands along coastlines, including that of the San Francisco Estuary (Greenberg et al. 2006b). They constitute a very distinctive biotic arena because of low overall plant diversity, wide variations in salinity, daily tidal inundation, and seasonal flooding by fresh water. In this setting, the terrestrial vertebrate fauna is relatively simple compared with that of the adjacent upland habitats. Many species that occur in tidal marshes are narrowly restricted endemic taxa, often with distinctive morphology, physiology, demographics, and behavior (Greenberg et al. 2006b). Consequently, tidal marshes along the Pacific coast are regarded as biodiversity hotspots. Their discontinuous distribution and isolated occurrences along the coast make them like an archipelago or string of pearls, a characteristic that makes faunal connections difficult and promotes isolation. In this context, Suisun Marsh sustains a remarkably diverse community of terrestrial animals, many of them coastal marsh endemics for which Suisun is a significant refuge. Historically the Marsh was considered a "veritable paradise" for birds and other wildlife (Skinner 1962), and in many respects it still functions in this manner (Olofson 2000; U.S. Fish and Wildlife Service [USFWS] 2009). In recent years, more than 180 species of birds, 45 mammals, and 15 reptiles and amphibians have been recorded from the broad range of habitats within the Marsh and in the adjacent transition zones from wetland to upland along its periphery. These numbers alone emphasize the importance of the Marsh ecosystem as a center for the conservation of regional terrestrial biodiversity. The Marsh and its adjoining upland habitats serve as critical terrain for many vertebrate species, including some with population levels that suggest impending extinction. Ecological policy and adaptive management planning for the future of the Marsh must encompass an integrated consideration of its wetland-dependent terrestrial vertebrates along with the obligate aquatic species. Indeed, Suisun Marsh is viewed as a place where restoration actions can compensate in part for losses of tidal marsh habitats elsewhere in the San Francisco Estuary, as well as being one of the few coastal wetlands where habitat conditions for sensitive terrestrial species can be improved (see chapter 1).

Because there is little current information about terrestrial invertebrates, they have not been emphasized in this book, with the exception of the conspicuous butterfly (Lepidoptera) fauna. Shapiro (1974) has recorded 40 species from the Marsh. Remarkably, most of these species are still present, despite widespread decline of native butterflies in the Central Valley (Shapiro 2009) and many other coastal wetlands (see box 6.2). We assume that management actions that enhance native terrestrial vertebrates can also favor and complement the conservation of native terrestrial invertebrates, as has been shown elsewhere (Del Guidice-Tuttle et al. 2011).

There are many challenges facing conservation of terrestrial vertebrates in the Marsh. Habitat alteration and fragmentation, urban development, alien species, contaminants, subsidence, and land- and water-use practices continue to threaten the Marsh, reducing its capacity to support a diverse terrestrial vertebrate community. The shortage of information on the status of most of these animals and on the cumulative effects of diverse threats complicates conservation efforts in the Marsh. Equally important, sea-level rise, climate change, and earthquakes will increasingly drive major changes to habitats in the Marsh in the next century (chapters 3, 8, and 9). In particular, all sea-level-rise scenarios predicted by NOAA (Parris et al. 2012), up to 2 m by 2100, will inevitably reduce Marsh terrestrial vertebrate habitat and populations.

This chapter briefly reviews the history, major stressors, and status and trends for important terrestrial vertebrates of the Marsh. It concludes with a brief discussion of the significance of the Marsh as habitat for key vertebrate species as major changes take place. The focus of this chapter is on species of special concern because they are rare or declining in the state, on charismatic species that are a focus of ecotourism, and on species that play important roles in the Marsh ecosystem (see table 6.1). Waterfowl merit a separate chapter (see chapter 5) because of their overall importance in Marsh history and management. Scientific names of native species mentioned in this chapter are given in table 6.1.

HISTORY

Pre-Euro-American Abundance

Prior to the arrival of Euro-American settlers, the Marsh was a large complex of diverse, interconnected, mostly tidal habitats (see chapter 2). High daily,

TABLE 6.1 Important terrestrial vertebrates of Suisun Marsh and their likely role in future Suisun Marsh management.

These species are listed under state or federal endangered species acts, are species of special concern, or are especially charismatic. "Marsh importance" reflects the likelihood of influencing management decisions in the Marsh and/or the importance of the species' role in the ecosystem. Ducks and geese are covered in chapter 5.

Common name	Scientific name	Status	Marsh importance	Notes
MAMMALS				
Suisun ornate shrew	Sorex ornatus sinuosus	SC/EN	H	Endemic to Marsh
Salt marsh harvest mouse	Reithrodontomys raviventris halicoetes	FE/SE	H	Endemic to region
California vole	Microtus californicus	CH	H	Keystone prey species; represents rodents in food webs
Brown rat	Rattus norvegicus	APC	M	Preys on eggs etc. of birds
American beaver	Castor canadensis	CH	H?	Role not well understood
Small-footed myotis bat	Myotis ciliolabrum melanorhinus	SC	L	Represents 14 species of bats, many SC
Tule elk	Cervus elaphus nannodes	CH	H	See box 6.1
Wild swine	Sus scrofus	APC	M	Growing importance
River otter	Lantra canadensis	CH	H	High abundance
Raccoon	Procyon lotor	CH	M	
Striped skunk	Mephitis mephitis	CH	M	Represents guild of small-medium predators
Coyote	Canis latrans	CH	H	Important predator
Red fox	Vulpes fulva	APC	L	Controlled by coyotes?
House cat	Felis catus	APC	M	Most important on marsh edges

ABBREVIATIONS: FE = federally endangered, FT = federally threatened, SE = state endangered, SC = state species of concern, SFP = state fully protected, CH = charismatic or important species for ecosystem function, EN = endemic species to region, APC = alien predator or competitor, H = high importance, M = moderate importance, and L = low importance.

(continued)

TABLE 6.1 (continued)

Common name	Scientific name	Status	Marsh importance	Notes
BIRDS				
Grebes				
Western grebe	*Aechmorphoris occidentalis*	CH	M	Represents migratory grebes
Pelicans and allies				
American white pelican	*Pelecanus occidentalis*	CH	H	Nonbreeding habitat
Double crested cormorant	*Phalacrocorax auritus*	CH	L	
Herons and allies				
American bittern	*Botaurus lentiginosus*	CH	M	
Least bittern	*Ixobrychus exilis*	SC	L	
Great blue heron	*Ardea herodias*	CH	M	
Great egret	*Ardea alba*	CH	H	
Snowy egret	*Egretta thula*	CH	M	
Black-crowned night heron	*Nycticorax nycticorax*	CH	M	
White-faced ibis	*Plegadis chihi*	CH	L	
Vultures				
Turkey vulture	*Cathartes aura*	CH	M	Marsh scavenger
Hawks, eagles, and falcons				
Northern harrier	*Circus cyaneus*	SC/CH	M	Abundant
White-tailed Kite	*Elanus leucurus*	ST/CH	M	
Red-tailed hawk	*Buteo jamaicensis*	CH	M	Represents diverse hawks
Swainson's hawk	*Buteo swainsoni*	ST/SFP/CH	L	
Ferruginous hawk	*Buteo regalis*	SC/CH	L	
Cooper's hawk	*Accipiter cooperii*	CH	L	Represents other non-special concern raptors
Bald eagle	*Haliaeetus leucocephalus*	FT/CH	L	
Golden eagle	*Aquila chrysaetos*	CH	M	Nests in uplands
Osprey	*Pandion haliaetus*	CH	L	
Peregrine falcon	*Falco peregrinus anatum*	SFP/CH	L	

Common name	Scientific name	Status	Importance	Notes
Prairie falcon	*Falco mexicanus*	SC/CH	L	
Rails and allies				
California black rail	*Laterallus jamaicensis coturniculus*	ST/SFP	H	
California clapper rail	*Rallus longirostris obsoletus*	SE/FE/EN	H	Regional endemic
Yellow rail	*Conturniop snoveboracensis*	SC	M	Edge of range
Shorebirds				
Marbled godwit	*Limosa fedoa*	SC/CH	M	Represents shorebirds
Long-billed curlew	*Numenius americanus*	SC/CH	L	
Western snowy plover	*Charadrius alexandrinus nivosus*	SC	L	Limited breeding
Gulls and terns				
Black tern	*Chlidonias niger*	CH	L	
California least tern	*Sterna antillarum browni*	FE/SE	M	Limited breeding
Owls				
Long eared owl	*Asio otus*	CH	L	
Short-eared owl	*Asio flammeus*	CH	M	
Western burrowing owl	*Athene cunicularia hypugea*	FE/SC/CH	M	Important upland bird
Other birds of interest				
Yellow-breasted chat	*Icteria virens*	CH	L	
Allen's hummingbird	*Selasphorus sasin*	SC	L	
Loggerhead shrike	*Lanius ludovicianus*	SC	H	
Tricolored blackbird	*Agelaius tricolor*	SC	M	Colonial nester
Suisun song sparrow	*Melospiza melodia maxillaris*	SC/EN	H	Endemic to Marsh
Salt Marsh common yellowthroat	*Geothlypis trichas sinuosa*	SC/EN	H	Regional endemic
Common raven	*Corvus corax*	CH	M	Common in upland areas
European starling	*Sturnus vulgaris*	APC	M?	Abundant but effects not known

ABBREVIATIONS: FE = federally endangered, FT = federally threatened, SE = state endangered, SC = state species of concern, SFP = state fully protected, CH = charismatic or important species for ecosystem function, EN = endemic species to region, APC = alien predator or competitor, H = high importance, M = moderate importance, and L = low importance.

(continued)

TABLE 6.1 *(continued)*

Common name	Scientific name	Status	Marsh importance	Notes
AMPHIBIANS				
California red-legged frog	*Rana draytonii*	SC/FT	L	Scarce in marsh
Pacific chorus frog	*Pseudacris regilla*	CH	L	Probably common but overlooked
Bullfrog	*Lithobates catesbeiana*	APC	M	Common in duck club ponds
Western spadefoot toad	*Spea hammondii*	SC	L	
California Tiger Salamander	*Ambystoma californiense*	FE	M	Upland species
REPTILES				
Giant garter snake	*Thamnophis couchi gigas*	ST/FT/EN	M	
Western skink	*Plestiodon skiltonianus*	FT	L	Upland species
Pacific pond turtle	*Emys marmorata marmorata*	SC/CH	M	

ABBREVIATIONS: FE = federally endangered, FT = federally threatened, SE = state endangered, SC = state species of concern, SFP = state fully protected, CH = charismatic or important species for ecosystem function, EN = endemic species to region, APC = alien predator or competitor, H = high importance, M = moderate importance, and L = low importance.

seasonal, and interannual variability in factors such as water level and salinity, as well as complex physical habitat diversity, resulted in an abundant and diverse terrestrial vertebrate community, with a mixture of resident and migratory species. The Marsh was also connected to other habitat areas throughout the region, especially the vast tidal marshes of the Sacramento–San Joaquin Delta, the Jepson Prairie region, and surrounding upland habitats and watersheds. This complex mosaic allowed many terrestrial vertebrates to use the region on a seasonal basis as migrants, as well as to develop interacting populations of resident species. It is likely that both numbers and habitats of terrestrial vertebrates were also manipulated by Native Americans who lived in and around the Marsh for the entire 6,000-year history of the Marsh (chapter 2). In particular, they cut and burned the tules on a regular basis to stimulate new stem growth for use in housing, food, basket making, and boats (Lightfoot and Parrish 2009). They also hunted many of the birds and mammals, no doubt affecting populations and changing behaviors.

Changes in the Past 200 Years

Over 200 years ago, Euro-American settlers began modifying the topography and hydrology of the Suisun region, ultimately reducing abundance of most vertebrates in the region. Many populations were severely reduced by overhunting, but major population declines were also directly associated with widespread habitat loss and degradation, as nearly 90% of the San Francisco Estuary's original tidal marsh became filled or modified (Olofson 2000).

Early Euro-American settlers exploited the abundant birds and mammals (see figure 6.1). The fur trade expanded rapidly in the 1800s because beavers, river otters, mink, foxes, weasels, and raccoons were plentiful (Skinner 1962). Big-game hunting was also popular and included tule elk, pronghorn antelope, black-tailed deer, grizzly bears, and black bears. Waterfowl and other marsh birds flocked throughout the Estuary and were shot by market hunters by the millions in the mid-1800s (Grinnell et al. 1918; Skinner 1962; chapter 5).

As settlers moved into the area, a large portion of tidal marshes in the Estuary, including Suisun Marsh, were converted into diked marshes, agricultural lands, and urban areas (see chapter 2). By the early 1900s, large animals such as grizzly bear, elk, and antelope were extirpated from the Suisun region. Additionally, there were huge losses of waterfowl and upland birds—most populations of these birds decreased by half in 40 years throughout the Estuary (Grinnell et al. 1918). By 1913, most shorebirds were protected from hunting because of their dramatic population declines (Skinner 1962). International treaties and stricter hunting regulations protected all waterfowl and migratory birds during the 20th century, as well as game and fur-bearing mammals, resulting in at least partial recoveries of many species. Additionally, duck hunting clubs became increasingly impor-

FIGURE 6.1. Herd of tule elk crossing Carquinez Strait in the 19th century. From William Heath Davis, *Seventy-Five Years in California* (San Francisco, 1929; courtesy of California Historical Society, CHS2010.345).

tant for providing waterfowl habitat. During this period, the Marsh evolved from natural marsh and agricultural fields to an area dominated by nontidal marshland managed for waterfowl hunting, with other terrestrial vertebrates adapting to the modified Marsh habitat (chapter 2).

Present Conditions

Modern Suisun Marsh and its adjacent upland habitat are highly altered and managed but still provide habitat for a diverse community of terrestrial species, although many of the species are in low abundances or transitory in their uses of Marsh and adjacent habitats. Changing conditions in the Marsh have shifted species assemblages and abundances. Some native species have declined dramatically or have been extirpated, while others have maintained robust populations. At the same time, alien species such as muskrat, starlings, and swine have invaded and established populations.

As conditions changed in the Marsh, there have been winners and losers among the terrestrial vertebrates. However, it is often hard to identify the exact mechanisms of species declines, especially for migratory populations, because external conditions may be more influential than those within the Marsh. For example, while waterfowl have benefited from intensive wetland management in the Marsh, numbers of pintail ducks have declined as a result of poor conditions in their breeding range (see chapter 5). Likewise, populations of other migratory birds, including shorebirds, songbirds, and raptors, may be more strongly affected by factors along migration routes than inside the Marsh.

A few resident mammals, such as black-tailed deer, raccoons, and introduced muskrats, appear to have increased initially as predator and competitor populations decreased (Skinner 1962). Some raptor populations likely increased as perching opportunities increased with the advent of power lines and tree planting, and as prey populations, especially rodents, flourished with conversion of marsh to grassland. Herons and egrets have also benefited from introduced eucalyptus trees (*Eucalyptus* spp.) that provide roosting and rookery habitat. By contrast, several native resident animals are at risk of extinction or extirpation, most notably salt marsh harvest mouse, California clapper rail, and California black rail. Some endemic species, such as Suisun song sparrow, saltmarsh common yellowthroat, and Suisun ornate shrew, have maintained populations but may also be at risk of extirpation as changes continue within and around the Marsh.

The habitat for terrestrial vertebrates will change dramatically in the Marsh in the next century as the result of multiple factors, most inevitably sea-level rise, climate change, and earthquakes. The likely extent of impacts from these changes is uncertain (chapters 3 and 8). It is important to note that these factors will be acting on terrestrial vertebrates that are already affected by other major stressors, both inside and outside the Marsh. In the next section, we discuss some of the major stressors that are currently affecting vertebrate populations: continued habitat alteration, habitat fragmentation, water and resource management actions, subsidence, invasions by alien species, contaminants, and disease.

MAJOR STRESSORS

Habitat Alteration

The habitat for terrestrial vertebrates in Suisun Marsh has undergone intensive alteration over the past 150 years. These changes have modified the physical template upon which the present distribution and abundance of terrestrial vertebrates depend. Major alterations include conversion of large areas of marsh to farm and grazing lands; construction of roads, railways, power lines, and dikes; and urban development. In the 1870s, when dike construction began, large tracts of marshland were drained and used for cattle grazing as well as for growing vegetables (Arnold 1996). By the 1920s, after agricultural failures, the predominant use of the Marsh was for cattle grazing and private duck hunting clubs (Olofson 2000). Additionally, to ease crop depredation in the Central Valley, the State of California purchased portions of the Marsh in 1927 to attract wintering waterfowl (Mall 1969).

Most alteration in recent decades has been to favor nontidal wetlands, in an effort to increase production of dabbling ducks and attract wintering waterfowl (chapter 5). About 210 km² (52,000 acres) of the Marsh's original 300 km² (74,000 acres) are now such managed nontidal wetlands (chapter 2; Arnold 1996). More than 400 km of dikes were constructed to separate interior areas from tidal sloughs and to manage freshwater flooding of duck club lands (Houghteling 1976). At present, diking and drainage systems provide fresh water when flooding is needed to attract waterfowl for hunting in fall–winter or for breeding in spring–summer. While this intense hunting-oriented management has favored many wetland species other than waterfowl (e.g., river otters), there have also been negative consequences from a conservation perspective. For example, invertebrate production in diked marsh land is often less than that of more natural tidal marsh with extensive mud flats; this can stress foraging vertebrates with high energy needs, such as Suisun shrew (WESCO 1986) and shorebirds, many of which are tidal mudflat specialists. The dike system also fragments habitat (Takekawa et al. 2006) and provides predators, such as house cats, access to prey that would otherwise be protected in natural tidal marsh with a complex island system (USFWS 2009). Finally, dikes require frequent maintenance that disturbs plant communities, potentially reducing their effectiveness as high-tide refuges for rails and rodents, as well as disturbing breeding and nesting habitat for birds (chapter 4; Takekawa et al. 2006).

Outside the Marsh proper, encroaching urbanization is a major cause of habitat alteration. It not only reduces habitat for wild vertebrates, but also reduces connectivity to other areas, potentially preventing some species, such as tule elk, from moving between the Marsh and upland areas. The Potrero Hills Landfill, located above the Marsh plain, likely affects vertebrates through upland habitat reduction and degradation. The problems are likely to increase if the landfill expands, as is planned. Urbanization also allows for increased invasions of wild areas by animals subsidized by humans, such as rats and cats. Additionally, human-caused fires can damage habitat. Another factor that accompanies habitat alteration, especially urbanization, is increased human disturbance. Such disturbance includes effects of recreation (e.g., hunting, jogging, fishing, boating, enjoying wildlife) and transportation (e.g., roads, boat wakes, loud noise). All types of disturbance can expose mammals and birds to predators (Takekawa et al. 2006), increase stress, reduce foraging time, and increase mortality through road kill.

Habitat Fragmentation and Connectivity

Fragmentation and isolation of populations exacerbates problems associated with habitat loss and alteration (Takekawa et al. 2006). In the Marsh, fragmented habitats resulted from many miles of dikes, power lines, canals, mosquito ditches, roads, railroads, and fences. Additionally, urbanization of the surrounding areas separates populations from other available open space. Fragmentation isolates populations, increasing extinction risk due to random population fluctuations and catastrophic events (Takekawa et al. 2006)

Fragmentation is especially likely to be a problem for species with low dispersal abilities, which may isolate small populations. The Suisun song sparrow may be adversely affected by fragmentation because this species flies only short distances from cover habitats (Larsen 1989; Scollon 1993), although the evidence is not clear on their maximum dispersal distance (Spautz and Nur 2008). Similarly, fragmentation and isolation of small populations is a concern for California clapper rail, as well as for California black rail (Olofson 2000; Spautz et al. 2006). Salt marsh harvest mouse and Suisun ornate shrew are also affected by fragmentation, because dispersal is difficult or impossible for them across large distances. Difficulties with dispersal may also affect other small mammals, reptiles, and amphibians that live in isolated marsh patches (Takekawa et al. 2006).

Connectivity between lowland marsh and upland areas provides refuges for species such as black rail from predators, tidal inundation, and dike failures (Evens and Page 1986). Improvement of such connectivity will be especially important as sea level rises and dikes fail. Tule elk, presently confined to Grizzly Island, exemplify the importance of habitat connectivity (box 6.1).

Water Management

For most of the Marsh, present management strategies focus on maintaining dikes and flooding regimes and increasing residence time of fresh water using the tidal gates on Montezuma Slough (chapters 3, 7, and 8), in order to promote favorable conditions for waterfowl and, secondarily, for other species. The ponds and ditches created by duck clubs are major foraging areas for river otters, herons, egrets, and other predators. Some of this managed habitat benefits locally adapted species such as Suisun song sparrow and giant garter snake (Olofson 2000). Maintaining the diverse and abundant terrestrial vertebrate community in Suisun Marsh is likely to result from maximizing habitat diversity along a salinity gradient created by the interaction of tides and freshwater inflow, as well as by a patchwork of land managed for different purposes.

It is worth noting that the present mosaic of terrestrial habitats in the Marsh is driven by human management efforts, even beyond the diked areas. Cattle grazing, construction of drainage ditches for mosquito control, use of herbicides and pesticides, dredging of sloughs, plant management (see chapter 4), disking of fields, and other practices have created habitats that favor a different suite of terrestrial vertebrates than what might be there naturally.

Alien Species

Alien species—especially those with invasive potential—are one of the most intractable problems in the Marsh, because once an alien species becomes estab-

BOX 6.1. TULE ELK IN SUISUN MARSH

Patrick Huber

The tule elk (*Cervus elaphus nannodes*) is a small subspecies endemic to central California. This charismatic herbivore originally occupied grasslands, wetlands, riparian forest, and coastal areas, often in large herds. These herds moved seasonally in search of fresh forage, mainly grasses, sedges, and forbs.

Gold mining and associated development, beginning in the mid-19th century, dramatically reduced the tule elk's populations. By the late 19th century there were fewer than 10, and perhaps as few as 2, remaining elk on one ranch in the San Joaquin Valley. Populations have since built up, and 8 elk were brought to Grizzly Island from the Tupman Reserve in the San Joaquin Valley in 1977. This herd is currently one of 21 throughout California, with a total statewide population estimated at 3,800.

The Grizzly Island herd has thrived since it was established. The population is controlled through hunting and through relocation of elk to augment existing herds or to establish new herds (at least four) in other locations in California. The elk generally use only a portion (~3,200 ha, or ~50%) of the total Grizzly Island area. However, there are occasional long-distance movements by individuals, either north from the Wildlife Area or south across Suisun Bay.

I have used landscape connectivity modeling to assess potential paths of movement for elk (see map 12 in color insert). Current land cover, roads, and land management were used to create a model to identify potential movement corridors between Grizzly Island, Yolo Bypass, and the Vaca Hills. My analysis

lished, it is very difficult to control. Alien invaders have both direct and indirect effects on native terrestrial vertebrates.

The most dramatic direct effect is predation. Important alien predators on native wildlife include red foxes, brown rats, house cats, and wild pigs. Red foxes and rats have been documented as serious threats to survival of California clapper rails (Harvey 1988), black-necked stilts, American avocets, and snowy egrets, as well as other birds (Harvey et al. 1992; Olofson 2000). Red fox trapping in other areas of the Estuary has positively affected bird populations and may become necessary for recovery of sensitive bird species in the Marsh (Olofson 2000; Takekawa et al. 2006). Rats are predators on rail eggs and chicks as well as other marsh-nesting birds (Harvey 1988). Feral and free-ranging house cats are less well studied in the Marsh but are known to kill birds and other small vertebrates they encounter, probably including Suisun ornate shrews and salt marsh harvest mice.

Feral pigs are more recent invaders. Although no studies have examined their effects on Marsh vertebrates, pigs are omnivorous, degrade habitat through their

has identified areas that, if managed properly, could be used by elk for access to currently unused habitat. For example, the Yolo Bypass possesses potential tule elk habitat, possibly enough to support dispersing elk. Further, it could serve as a gateway to other potential habitat in other parts of the Sacramento Valley. The bypass is likely to be an important component of any future network of tule elk populations if greater restoration of the subspecies is pursued.

A herd of tule elk that would benefit from interactions with Marsh elk lives in the Coast Range, north of the Vaca Hills; connectivity between this herd and the Grizzly Island herd will depend on enhancement of habitat between Vacaville and Fairfield to create a corridor for movement. There is currently a narrow gap between the cities in the vicinity of Lagoon Valley; however, the I-80 freeway bisects this linkage. An elk crossing (either an under- or overpass) would be needed for elk to successfully move between the two herds. Like deer and many other large herbivores, elk will use well-designed crossings for seasonal movement. These linkages merge in the Jepson Prairie region, a key node in a potential future network. Establishment of this conservation network could also provide movement potential for other species, linking Suisun Marsh to other important habitats throughout the Central Valley.

As sea level rises or as elk populations expand, or both, additional habitat could be created for elk in the Portrero Hills. This presumably would require replacing grazing by cattle with grazing by elk, and recreating a more complex habitat mosaic (e.g., expanded riparian habitat) and more diverse plant assemblages, dominated by native plants. Such a change in management would benefit other upland species as well, such as burrowing owls and badgers.

rooting behavior, and devour ground animals such as nesting birds (e.g., rails) and reptiles. Suisun Marsh may be especially at risk from feral pigs because wetland habitats are highly susceptible to damage from their rooting behavior (Choquenot et al. 1996).

In a more subtle manner, alien plants can alter habitat structure in the Marsh (see chapter 4), affecting the animals that depend on it. For example, hybrids between native and alien cordgrass (*Spartina*) species (Daehler et al. 1997) in San Francisco Bay marshes grow in denser stands than the native species and invades of mudflats, reducing the ability of marshlands to sustain cordgrass-dependent species such as the California clapper rail, as well as foraging shorebirds. The cordgrass invasion is likely to become a problem in Suisun Marsh as sea level rises and creates saltier edge habitats.

However, some alien plants, such as eucalyptus trees and giant reed (*Arundo donax*), provide perching and roosting habitat for many predatory birds, including herons, owls, and various raptors.

Contaminants

Research in other parts of the Estuary has demonstrated impacts of contaminants that are likely occurring in the Marsh as well (Schwarzbach et al. 2006). Contaminants such as arsenic, chromium, copper, nickel, DDT, chlordanes, dieldrin, dioxins, furans, and polybrominated diphenyl ethers (PBDEs) are regarded as "priority pollutants" for the Estuary because of their known negative effects on terrestrial vertebrates, especially birds (San Francisco Estuary Institute 2003). Effects include reproductive failure, decreased hatch success, and disrupted endocrine function.

Mercury-laden sediments deposited in the Marsh over many years, the legacy of gold mining in the upper watershed, are an ongoing concern. Restoration and management projects in the Marsh may depend on disturbing existing sediment (dredging) or filling other areas with new sediment. Sediment in the Marsh is often laden with mercury, selenium, and polychlorinated biphenyls (PCBs). A study in nearby San Pablo Bay demonstrated that when mercury is released from sediments during the draining of duck club ponds, it is transformed by bacteria into methyl mercury, which can be toxic to birds and other predators (Marvin-DiPasquale et al. 2003). Similar results have been found for the draining of duck clubs ponds in Suisun Marsh (S. Siegel, personal communication, 2013). Exposure to mercury can affect vertebrate reproduction (Wiener and Spry 1996; Wolfe et al. 1998) and damage the central nervous systems of fishes, birds, and mammals (Wolfe et al. 1998; Wiener et al. 2003). Additionally, mercury, selenium, and PCBs are known to affect egg viability of clapper rails (Schwarzbach and Adelsbach 2002; Novak et al. 2005; Schwarzbach et al. 2006) and may affect black rails as well (USFWS 2009).

Disease

West Nile virus, avian cholera, St. Louis encephalitis, and avian botulism are four diseases that may affect bird and other wildlife populations in Suisun Marsh. West Nile virus, transmitted by mosquitoes, was detected in 2004 in birds of the Suisun region. The highest prevalence was found in the crows, jays, and magpies (Corvidae). Cormorants, shorebirds, and song sparrows were found to be hosts as well. An indirect effect of increases in mosquito-borne diseases is the increase in mosquito control efforts, which include pesticide application, introduction of non-native vector control species (e.g., mosquitofish), and draining of some marsh areas.

STATUS AND TRENDS

Birds

Although habitat in the Marsh has been highly altered, 180 species of birds have been recorded from the Suisun region. While many are upland birds or tropical migrants with rare occurrences in the Marsh, nearly every regional wetland bird species is represented. Suisun Marsh is important habitat for a number of Bird Species of Special Concern (BSSC; Shuford and Gardali 2008) or state listed species (Spautz et al. 2012). Here, we focus mainly on marsh-dependent birds, including species of special concern, fish-eating birds, shorebirds, and raptors. Their protection is often important for management of different Marsh areas, especially public lands. Ducks and geese are covered in chapter 5. Important species of special concern include two species of rails, Suisun song sparrow, saltmarsh common yellowthroat, tricolored blackbird, and loggerhead shrike (see table 6.1).

California Black Rail.   The black rail is a tiny, secretive bird that forages among tidal marsh plants and is most often detected through its nocturnal calling. It favors, especially for breeding, dense vegetation with freshwater influence, but with infrequent flooding (Evens et al. 1989, 1991; Evens and Nur 2002). Black rails will use more open, restored marshlands for foraging in the nonbreeding season if such habitat is close to patches of denser habitat (Olofson 2000). Although this species is rarely seen and is listed by the state as a threatened species, recent studies indicate that black rail populations may be increasing at Rush Ranch and Suisun Bay edge marshes (Spautz et al. 2012). The rails are driven from their secret places by the flooding of high tides. They seek refuges on high ground, even dikes. Here, if plant cover is insufficient, they are easy (if small) prey for great egrets, northern harriers, foxes, and house cats (Evens and Page 1986). Dikes will likely become increasingly important as refuges as a consequence of sea-level rise in the future, when marsh vegetation is reduced and inundated more frequently.

California Clapper Rail.   Elusive California clapper rails thrive in large areas of high tidal marsh, with tall and dense vegetative cover of silverweed (*Potentilla anserina*), tules (*Schoenoplectus* spp.), and cattails (*Typha* spp.) (Foin et al. 1997; California Department of Fish and Game 2004). They move about in the intricate channel networks of tidal marshes that provide escape routes from predators and support abundant invertebrate populations for food (Olofson 2000). The once abundant California clapper rail, now federally listed as an endangered species (Olofson 2000), is found throughout the Marsh (Gill 1979; Olofson 2000). However, the Marsh population may have declined in recent years (Spautz et al. 2012). Like the smaller black rails, these birds suffer from a lack of adequate refuges to escape predators during high tides, lack of connectivity among fragmented populations, and habitats that are highly altered, such as duck hunting areas that are drained in summer and planted with alien vegetation. Because the rail's main defense against predators is hiding, all life stages from eggs to adults are quickly devoured if cover is inadequate. Raptors were found to kill 64% of rails when high vegetative cover was sparse, primarily during winter (Albertson

1995). Power-line poles and exotic trees provide additional perching opportunities for raptors to prey on rails, especially when high tides chase the rails from their hiding places. While plant-covered dikes can provide refuges from high tides, they also increase access for predators such as foxes, raccoons, skunks, feral cats, and rats. Nonnative red foxes have been implicated as major predators on rails (Foerster et al. 1990), but they are less of a problem in Suisun Marsh because coyotes suppress the fox population (Olofson 2000). In the Marsh, native river otter populations are high, and they may be opportunistic predators on rails as well.

Habitat fragmentation is believed to be a major factor inhibiting clapper rail recovery because small, isolated populations combined with high predator pressure and low dispersal rates make population persistence exceptionally difficult (Foerster et al. 1990; Olofson 2000). Additionally, these small populations are more likely to become locally extinct from stochastic events. Recreating large blocks of tidal marsh with high plant density, complex tidal channels, and high-tide islands is a priority for the survival of this species (Olofson 2000) and would likely benefit other marsh-dependent species as well.

Suisun Song Sparrow.   The characteristic song of this small brown bird helps to define Suisun Marsh. Its entire population is restricted to Suisun Marsh, so it is a BSSC. It is associated with high-quality tidal marsh because it requires tall, dense vegetation for calling and nesting (Larsen 1989; Scollon 1993; Marshall and Dedrick 1994). However, it will also use lower-quality habitat in diked wetlands (Olofson 2000; Spautz and Nur 2008). For example, Resh (2001) found that drainage ditches created for mosquito control increased song sparrow numbers. Adult song sparrows occupy territories for life in dense vegetation (Spautz and Nur 2008), and Larsen (1989) suggested that they fly only short distances from such areas (but see Spautz and Nur 2008). Their territories are typically associated with vegetation in fresh or brackish water, so their populations could be affected by increasing salinities that change plant species composition (Scollon 1993; Spautz and Nur 2008).

As with all small marsh birds, predation is a major threat to song sparrow survival. Predators include aliens such as Norway rats, red foxes, and feral cats, as well as native garter snakes (*Thamnophis* sp.), American crows, common ravens, short-eared owls, and northern harriers (Olofson 2000; Spautz and Nur 2008). Greenberg et al. (2006a) found a mean nest failure of 90% for song sparrows in Suisun Marsh, largely attributed to predation. Nesting was monitored at Rush Ranch from 1996 to 2005, and low nest survival rates indicated that song sparrow populations were not likely to persist (Spautz et al. 2012).

Saltmarsh Common Yellowthroat.   The saltmarsh common yellowthroat, also referred to as "San Francisco yellowthroat," is an energetic songbird endemic to

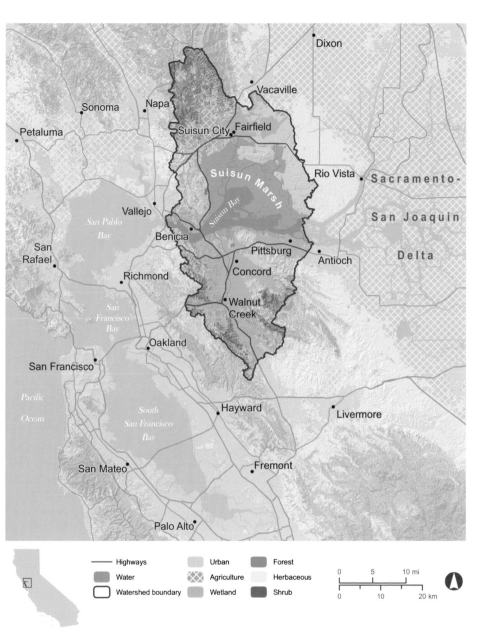

MAP 1. The San Francisco Estuary, showing the position of Suisun Marsh in relation to the Delta and San Francisco Bay, as well as regional land cover types (Gesch et al. 2002; CalAtlas 2012; U.S. Geological Survey 2013).

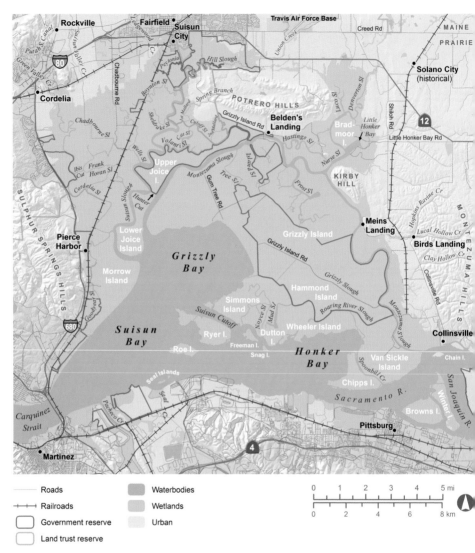

MAP 2. Suisun Marsh place names and notable features (Gesch et al. 2002; CalAtlas 2012; Contra Costa County 2013; Solano County 2013).

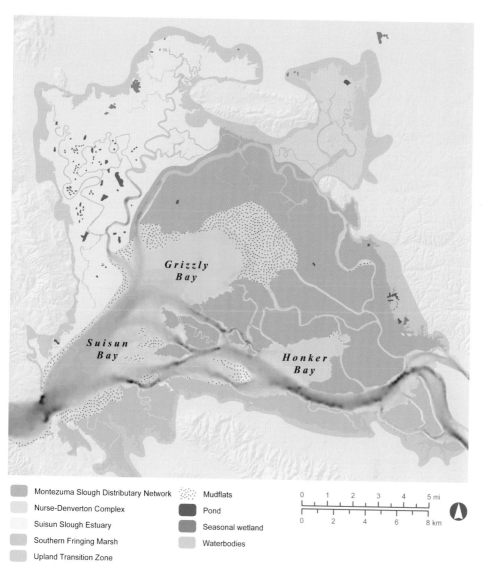

Legend:

- Montezuma Slough Distributary Network
- Nurse-Denverton Complex
- Suisun Slough Estuary
- Southern Fringing Marsh
- Upland Transition Zone
- Mudflats
- Pond
- Seasonal wetland
- Waterbodies

0 1 2 3 4 5 mi

0 2 4 6 8 km

Grizzly Bay

Suisun Bay

Honker Bay

MAP 3. Historical functional subregions of Suisun Marsh, late 1800s. Sloughs, landforms, and bathymetry are from measurements taken during 1856–67 (Bache 1872). Ponds and wetlands are from topographic maps surveyed during 1896–1907 (U.S. Geological Survey 1896, 1918a, 1918b). Topography is contemporary (Gesch et al. 2002).

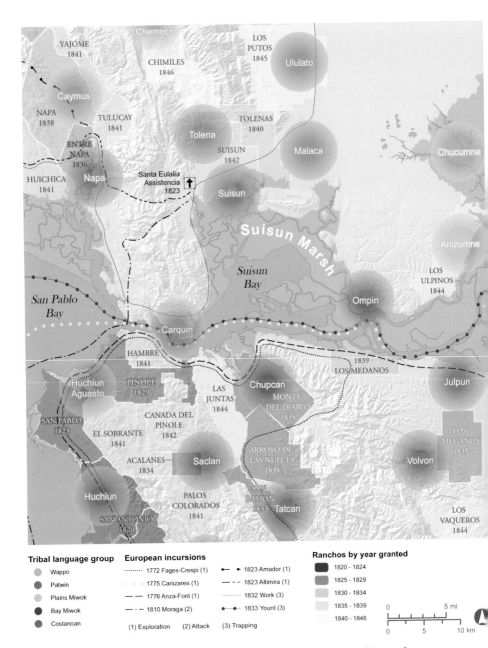

MAP 4. Shifting landscape pressures, 1772–1846: a representation of known human presence, intended for conceptual illustration. Circles show approximate centers of areas inhabited by tribal groups; colors represent Indian language groups. Also shown are early explorers' paths and Mexican land grants. (Geospatial data sources: Eldredge 1909; Camp and Yount 1923; Maloney and Work 1943; Shumway 1988; U.S. Bureau of Land Management 1993; Milliken 1995; Gesch et al. 2002; Bowen 2009; San Francisco Estuary Institute 2012; Whipple et al. 2012.)

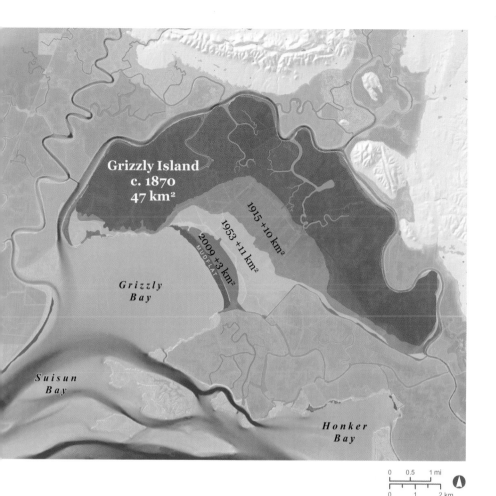

Grizzly Island
c. 1870
47 km²

1915 +10 km²

1953 +11 km²

2009 +3 km²

MUDFLAT

Grizzly
Bay

Suisun
Bay

Honker
Bay

0   0.5   1 mi
0   1   2 km

MAP 5. Grizzly Island reclamation of Gold Rush sediments, 1870–2007. Successive dikes
built after the Swamp Land Reclamation Act captured gold-mining sediment moving
through the estuary. Dates refer to map publication, not levee installation. Present-day
topography and bathymetry are shown. (Geospatial data sources: U.S. Geological Survey
1949 [1980], 1953a [1980], 1953b [1980]; Metsker 1953; U.S. Bureau of Land Management
1993; Gesch et al. 2002; California Department of Water Resources 2007; Boul et al.
2009; Foxgrover et al. 2012.)

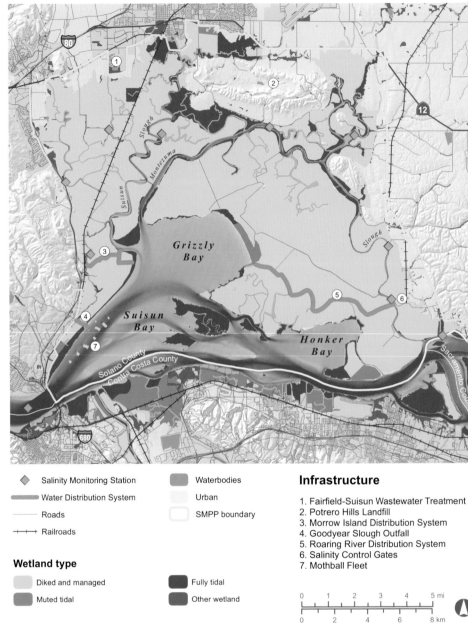

Salinity Monitoring Station

Water Distribution System

Roads

Railroads

Wetland type

Diked and managed

Muted tidal

Waterbodies

Urban

SMPP boundary

Fully tidal

Other wetland

Infrastructure

1. Fairfield-Suisun Wastewater Treatment
2. Potrero Hills Landfill
3. Morrow Island Distribution System
4. Goodyear Slough Outfall
5. Roaring River Distribution System
6. Salinity Control Gates
7. Mothball Fleet

| 0 | 1 | 2 | 3 | 4 | 5 mi |
| 0 | | 2 | 4 | | 6 | 8 km |

MAP 6. Hydrologic infrastructure and wetland types in Suisun Marsh today. (Geospatial data sources: Gesch et al. 2002; CalAtlas 2012; Foxgrover et al. 2012; San Francisco Estuary Institute 2012; California Department of Water Resources 2013; Contra Costa County 2013; Solano County 2013.)

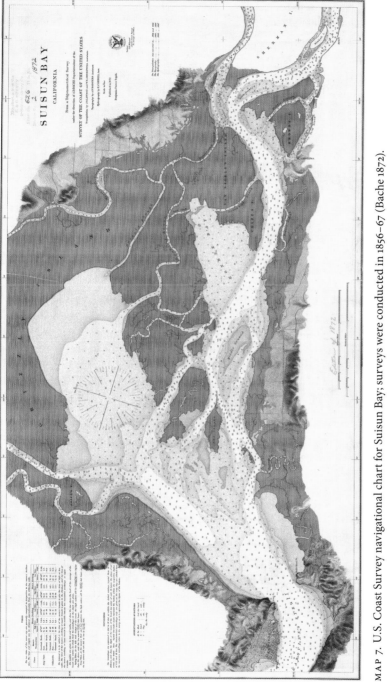

MAP 7. U.S. Coast Survey navigational chart for Suisun Bay; surveys were conducted in 1856–67 (Bache 1872).

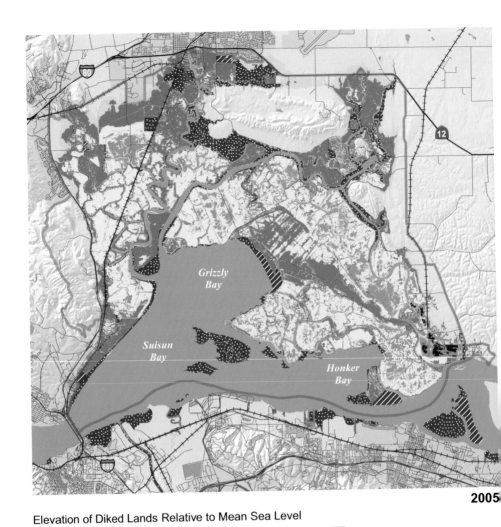

2005

Elevation of Diked Lands Relative to Mean Sea Level

- +6.5 to +11.5 ft (above intertidal)
- +1.5 to +6.5 ft (intertidal)
- -1.5 to +1.5 ft (shallow subtidal)
- -4.5 to -1.5 ft (medium subtidal)
- < -4.5 ft (deep subtidal)
- Open Water (unclassified)

Other Features

- Muted Tidal
- Fully Tidal
- SMPP boundary
- Roads
- Railroads
- Urban

0 1 2 3 4 5 mi

0 2 4 6 8 km

MAPS 8 and 9. Elevation of Suisun Marsh in relation to present sea level (map 8) and in relation to expected 1.4-m sea-level rise by 2100 (map 9). (Geospatial data sources: Gesch et al. 2002; California Department of Water Resources 2007; CalAtlas 2012; Foxgrover et al. 2012; San Francisco Estuary Institute 2012; California Department of Water Resources 2013; Contra Costa County 2013; Solano County 2013.)

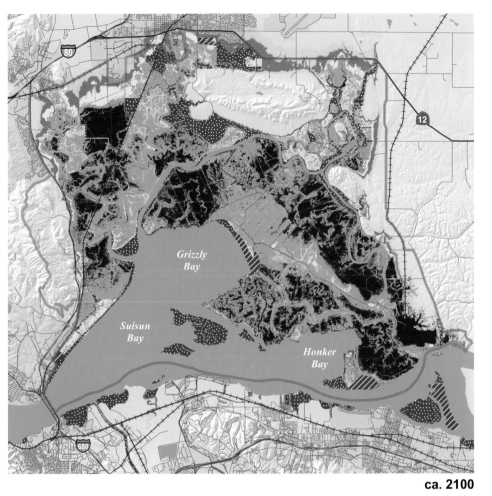

ca. 2100

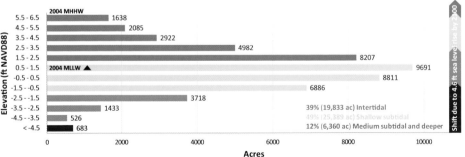

Distribution of existing elevations of diked lands in Suisun Marsh. The median elevation of diked lands in the Marsh is about mean lower low water (MLLW, 1.5 ft); 39% of the diked land is above MLLW, 49% is between zero elevation and −3 ft, and 12% is more than 3 ft below MLLW (for spatial distribution of these diked land elevations, see map 8 in color insert). Arrow on right shows how the areal distribution of tidal lands would shift with sea-level rise by 2100.

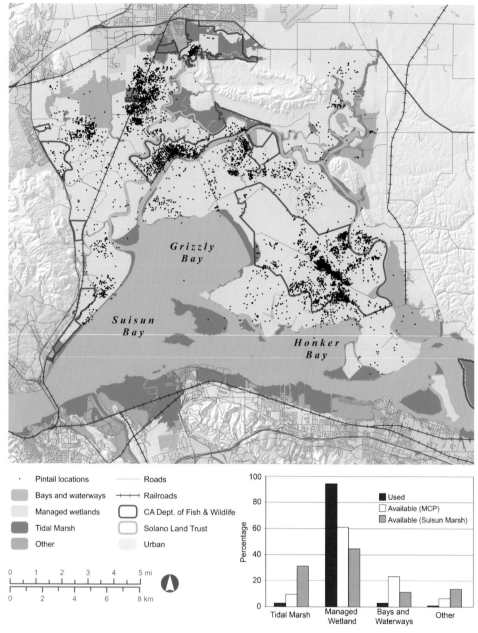

Legend:
- · Pintail locations
- Bays and waterways
- Managed wetlands
- Tidal Marsh
- Other
- Roads
- Railroads
- CA Dept. of Fish & Wildlife
- Solano Land Trust
- Urban

Graph legend:
- Used
- Available (MCP)
- Available (Suisun Marsh)

Graph y-axis: Percentage (0–100)
Graph x-axis categories: Tidal Marsh, Managed Wetland, Bays and Waterways, Other

MAP 10. Locations (*N* = 7,825) of 215 radiomarked female northern pintail in Suisun Marsh habitats from August to February during winters 1990–93. The graph shows habitat use by radiomarked pintails in relation to habitat availability in the Marsh. Used habitats are represented by the percentage of radiomarked pintail locations found within each habitat type. Habitat availability was estimated two ways, as those habitats available within (1) the Suisun Marsh Basin boundary as pictured in figure 1.1 and (2) the minimum convex polygon (MCP) depicting the home range of the entire population of 215 radiomarked pintails. Habitats were categorized as tidal marsh, managed wetlands, bays and waterways, and other habitat (which included uplands and grazed lands). (Waterfowl data provided by authors. Geospatial data sources: Gesch et al. 2002; CalAtlas 2012; San Francisco Estuary Institute 2012; Contra Costa County 2013; Solano County 2013.)

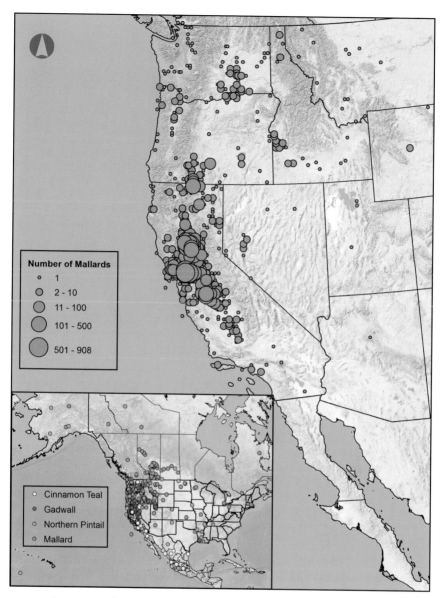

MAP 11. Location and abundance of ducks captured and banded in Suisun Marsh during late summer (May–September) and recovered (N = 9,368) since 1932 in North America. The main map shows recovered mallard (orange) in the western United States, and the inset map shows recovered mallard (orange; N = 8,367), northern pintail (green; N = 670), gadwall (blue; N = 246), and cinnamon teal (yellow; N = 85) in North America. Ducks were banded within the Suisun Marsh by the California Department of Fish and Wildlife and California Waterfowl Association, and band recoveries were thereafter managed by the U.S. Geological Survey's Bird Banding Lab.

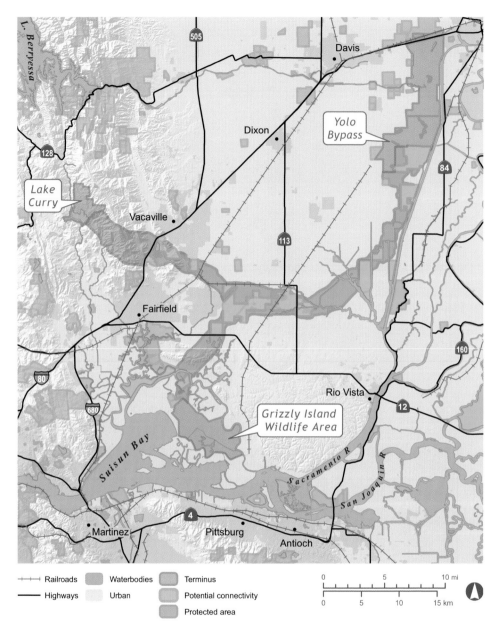

MAP 12. Potential pathways of movement for tule elk between Grizzly Island, Yolo Bypass, and the Vaca Hills. The program used chooses the best routes from Grizzly Island to other potential refuge (terminus) areas.

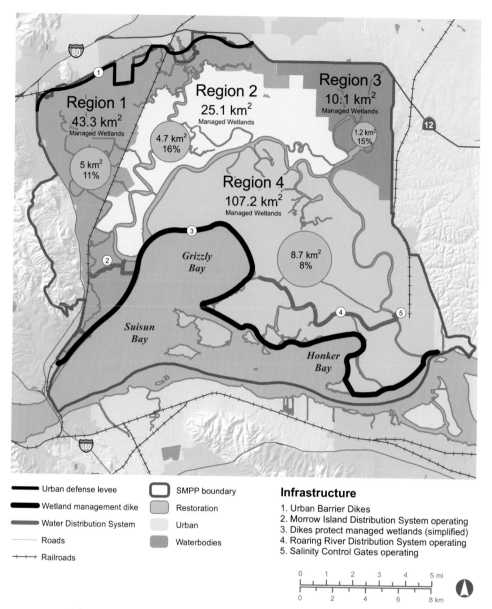

MAP 13. The Fortress Marsh scenario. The regions and amount of proposed restored marsh area (circles) follow the Suisun Marsh Plan of Protection (2010). This scenario envisions Suisun Marsh being maintained as close to its present-day state as possible (Gesch et al. 2002; San Francisco Bay Conservation and Development Commission 2010; CalAtlas 2012; Contra Costa County 2013; Solano County 2013).

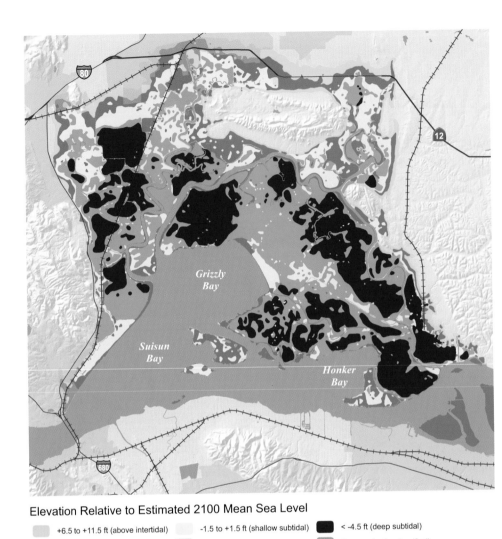

Elevation Relative to Estimated 2100 Mean Sea Level

+6.5 to +11.5 ft (above intertidal)

+1.5 to +6.5 ft (intertidal)

-1.5 to +1.5 ft (shallow subtidal)

-1.5 to -4.5 ft (medium subtidal)

< -4.5 ft (deep subtidal)

Open water (unclassified)

Other Features

——— Roads

+++++ Railroads

Urban

0 1 2 3 4 5 mi

0 2 4 6 8 km

MAP 14. The Flooded Marsh scenario. Without proactive efforts to reverse subsidence, much of the Marsh may be inundated in the future (California Department of Water Resources 2007; CalAtlas 2012; Contra Costa County 2013; Solano County 2013).

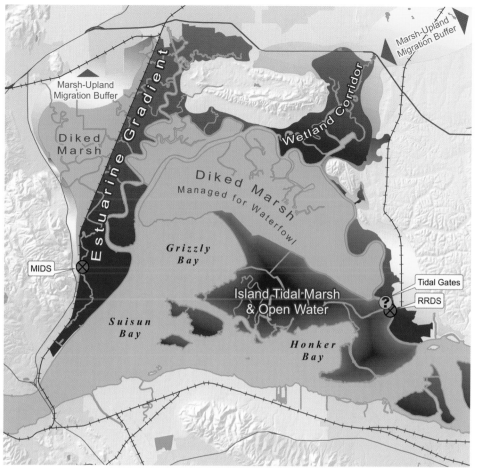

Roads

Railroads

Salinity Control Gates repurposed or non-operational

Morrow Island and Roaring River Distribution Systems non-operational

Urban

Waterbodies

| 0 | 1 | 2 | 3 | 4 | 5 mi |
| 0 | 2 | 4 | 6 | 8 km |

MAP 15. The Planned Accommodation Marsh scenario. This would be a mixed-use marsh meeting many objectives (CalAtlas 2012; Contra Costa County 2013; Solano County 2013).

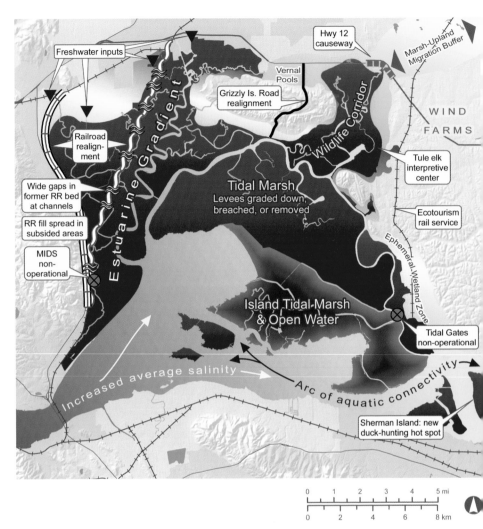

Freshwater inputs

Hwy 12
causeway

Marsh-Upland
Migration Buffer

Vernal
Pools

Grizzly Is. Road
realignment

WIND

FARMS

Railroad
realign-
ment

Wildlife Corridor

Tule elk
interpretive
center

Wide gaps in
former RR bed
at channels

Estuarine Gradient

Tidal Marsh
Levees graded down;
breached, or removed

RR fill spread in
subsided areas

Ecotourism
rail service

MIDS
non-
operational

Ephemeral Wetland Zone

Island Tidal Marsh
& Open Water

Tidal Gates
non-operational

Increased average salinity

Arc of aquatic connectivity

Sherman Island: new
duck-hunting hot spot

0   1   2   3   4   5 mi

0   2   4   6   8 km

MAP 16. The EcoMarsh scenario. Increased habitat connectivity restores regional
function (CalAtlas 2012; Contra Costa County 2013; Solano County 2013).

the San Francisco Estuary and is a BSSC (Gardali and Evens 2008). Its highest densities anywhere are in Suisun Marsh, where the population seems to be stable (Liu et al. 2007). Yellowthroats are year-round residents of tidal marshes but may also occur in other wetland habitats, depending on time of year (Nur et al. 1997; Olofson 2000). Nur et al. (1997) found their greatest abundance to be in areas that contained extensive tidal channels; they were negatively associated with pickleweed but positively associated with tules, peppergrass, and common cattail. Generally, these habitat requirements mean they are not found in abundance in duck club areas, especially those that are drained by June. Threats to persistence include habitat loss, predation from the standard array of predators, and brood parasitism by brown-headed cowbirds.

Tricolored Blackbird.   A BSSC, the tricolored blackbird has declined largely as a result of habitat loss, especially loss of habitats that support their colonial breeding (Beedy 2008). In Suisun Marsh there is foraging, breeding, and roosting habitat for them. A breeding colony of approximately 200 birds was observed at Rush Ranch in 2008 (U.S. Bureau of Reclamation [USBR] et al. 2011). Current threats to these birds in the Marsh include habitat loss, land management practices that disturb the soil, predation, and contaminants (Beedy 2008). Black-crowned night-herons, common ravens, and coyotes have been documented preying on entire or large portions of tricolored blackbird colonies in other parts of the Estuary (Beedy 2008). Contaminants such as mosquito-abatement oil, herbicides, and selenium have all caused major losses of birds in the past but need to be assessed as to their impact on blackbirds in the Marsh (Beedy 2008).

Loggerhead Shrike.   This distinctive, black-masked bird is a BSSC (Humple 2008) but is fairly common in the Marsh, especially along the edges in the transition zone from marsh to upland habitat. The causes of its decline are poorly understood in California, but habitat loss, agricultural practices, fire, contaminants, poor wintering habitat, and vehicle collisions are implicated as possibilities (Humple 2008). As the species declines in the state, the population in Suisun Marsh becomes increasingly important. It will be a good indicator of success of maintaining transitional habitats as sea level rises.

Fish-eating Birds.   Suisun Marsh is an important rearing area for diverse fishes (chapter 7), so it supports substantial populations of fish-eating birds that catch fish in channels and ponds on both sides of the dikes. Of particular interest are the conspicuous herons and egrets that are present year around and breed in the Marsh. Some of these birds are in decline and have special conservation status (table 6.1), including black-crowned night heron (federal sensitive species), great blue heron (state sensitive), and least bittern (BSSC). Other, more abundant, species include snowy egret, great egret, and American bittern. Large breeding colonies of herons are found in the Marsh, many in eucalyptus trees on duck clubs (Kelly et al. 2007). The Marsh is especially important for great egrets; the hundreds of breeding pairs make up about 35% of the Bay Area population (Kelly et al. 1993). Productive great egret nesting colonies are associated with larger sloughs with abundant fish (Kelly et al. 1993). Kelly et al. (2007) reported regional declines in reproductive success of great blue herons, black-crowned night-herons, and snowy egrets. Declines were concomitant with increases in the abundance of common raven and American crow, known nest predators, but causality has not been determined (Kelly et al. 2007). Herons and egrets are known predators of other terrestrial vertebrates. In addition to fish, they prey on vulnerable birds such as black rails (Evens and Page 1986), as well as on reptiles and amphibians such as garter snakes, alligator lizards, California toads, and Pacific chorus frogs (accounts in Olofson 2000).

Other fish-eating birds in the Marsh include western grebes, double-crested cormorants, American white pelicans, and assorted gulls and terns. White pelicans (BSSC) have only recently begun summering in the Marsh, mainly in diked wetlands (figure 6.2; see www.audubon.org/bird/iba). California least terns have recently (2006 and onward) been documented successfully nesting in the Montezuma Wetland restoration site (USBR et al. 2011).

Shorebirds. Many shorebirds use the Marsh, especially during winter or during migratory periods, foraging on tidal mudflats and in shallow ponds (USFWS 2009). They include snipe, dowitchers, plovers, curlews, willets, avocets, godwits, stilts, sandpipers, yellowlegs, phalaropes, and killdeer, but their numbers seem to be greatly reduced from former levels. Overhunting took a heavy toll in the 1800s and 1900s, but the present low levels are the result of loss of mudflats and other foraging habitats (Skinner 1962). Of these birds, only the western snowy plover is listed (federally, as a threatened species). It has been documented breeding in the Marsh (Olofson 2000).

Raptors. Raptors probably experienced increased abundance in the Marsh as perching opportunities increased with the introduction of power lines, buildings, and eucalyptus trees. Particularly abundant as foragers over marshlands are white-tailed kites, northern harriers, red-tailed hawks, and American kestrels. Northern harriers (BSSC) and white-tailed kites (California fully protected species) are both year-round residents within the Marsh (USBR et al. 2011).

Upland areas are used for breeding and roosting by golden eagles, short-eared owls, and great horned owls (USBR et al. 2011). As development continues in Solano County, preserving upland areas around the Marsh will become increasingly important for these birds of prey. For example, the only two known nesting

FIGURE 6.2. White pelicans take flight in Suisun Marsh (photo by A. Manfree).

sites for golden eagles in Solano County are in the Potrero Hills (California Department of Water Resources 2011).

Of particular concern are short-eared owls (BSSC) that winter in the Marsh in large numbers; they are also the only resident population in the region (Roberson 2008). Nests are located in the upland areas during September through April (USBR et al. 2011). Threats to survival include habitat loss and degradation. For example, cattle grazing can destroy cover needed for ground nesting, as well as destroy nests and eggs. In addition, the owls are affected by ground predators (e.g., alien red foxes) and diseases such as west Nile virus (Roberson 2008).

Burrowing owls (BSSC) are permanent residents in upland areas (USBR et al. 2011), in grasslands and agricultural fields, wherever there are burrows of ground squirrels that the owls require for nesting and roosting. In the Suisun region, numbers of breeding owls have apparently decreased for unknown reasons (Gervais et al. 2008).

Mammals

Forty-five species of mammals may occur in the Marsh and its peripheral habitats at one time or another, but it is likely that only about a dozen (not counting bats) are common in the Marsh or have a strong dependence on the habitats it provides. Increasing human populations and associated habitat loss over the past 150 years have negatively affected many mammals of the Marsh. Large species such as tule elk, pronghorn antelope, black bears, and grizzly bears were all extirpated throughout the Marsh region. Only tule elk have been reintroduced (box 6.1). The larger mammals that remain represent a suite of alien and native species, such as alien feral pigs, house cats, red foxes, and domestic cattle and native coyotes, striped skunks, badgers, and raccoons. It is likely that predation on small native mammals and birds of the Marsh is especially high because (1) the urban fringes support high densities of alien predators, such as house cats; and (2) dikes and other habitat changes allow these predators more access to interior habitats that might otherwise be refuges from mammalian predation. However, it is also likely that mammalian predators have always been important in regulating populations of other marsh vertebrates, including nesting waterfowl. For example, Ackerman (2002) documented the complex interactions related to predation by mammalian predators on waterfowl nests and rodents in the Marsh; when rodents are abundant (especially California voles) there is less predation on duck nests. Both rodent abundance and waterfowl nesting success depend in part on the way habitat for them is managed. Likewise, predation on small mammals and birds by red foxes may be reduced because coyotes reduce fox populations. However, most predator–prey interactions in the Marsh are poorly understood, as reflected in what is known about four important mammal species affected by Marsh management: salt marsh harvest mouse, Suisun shrew, American beaver, and river otter.

Salt Marsh Harvest Mouse.   Populations of this federal and state endangered species have been diminished by large-scale loss and alteration of tidal marsh (Shellhammer 1989). Sea-level rise will doubtlessly continue this trend as suitable tidal marsh habitat is inundated. Today, populations are found mainly in the middle and high tidal zones of Suisun Marsh (USBR et al. 2011) as well as in diked wetlands (Sustaita et al. 2011). This species needs access to high-tide refuges for cover and escape from predators (Shellhammer 1989). This can either be dikes and upland refuges or tall vegetation (K. R. Smith, personal communication, 2013). Recurrent shallow flooding, access to tidal refuges, and habitat connectivity are noted as important factors in recovery of this species throughout the Estuary. Much emphasis has been put on restoring tidal wetlands for recovery of mouse populations. Bias and Morrison (2006) have suggested that competitive interactions with house mice and voles may threaten mouse populations, although Sustaita et al. (2011) found no evidence of their negative effects on other rodents in Suisun Marsh. Sustaita et al. (2011) also found that different population parameters, such as density and postwinter persistence, varied among wetland habitats in the Marsh, but that there was no significant difference in reproductive potential and short-term persistence of mouse populations in tidal and diked wetlands, although both habitats are used. However, further study is needed to determine how to make intensely managed diked wetlands, such as those managed for waterfowl in the Marsh, support high densities of salt marsh harvest mouse (K. R. Smith, personal communication, 2013). Invasion by Algerian sea lavender and its displacement of native marsh vegetation may be another future problem that could significantly reduce mouse habitat.

Suisun Ornate Shrew. The hyperactive Suisun ornate shrew is a regional endemic subspecies that is currently limited to remaining tidal salt and brackish marshes along Suisun and San Pablo Bays (WESCO 1986). Like all shrews, it has high energetic needs (Newman 1970), so it prefers foraging in tidally influenced wet soils where amphipods and other invertebrates are abundant (WESCO 1986; Hays 1990). The historical switch from highly productive tidal marsh to mostly diked nontidal marsh may have negatively affected Suisun shrew populations because reduced invertebrate densities did not provide adequate food (WESCO 1986). Surprisingly, these marsh shrews can die if their fur becomes wet and cold, which makes them susceptible to flooding (Hays 1990). When relatively taller plants with dense foliage are limited, it is difficult for shrews to avoid getting wet. During extreme high tides, mortality can occur. Connectivity also is essential for survival of Suisun ornate shrews because they have difficulty dispersing across large distances. Overall, restored tidal marsh and connectivity among remnant populations are important for this species' survival, especially in relation to sea-level rise.

River Otter. The river otter is one of the most charismatic species in the Marsh because of its abundance and visibility. Grenfell (1974) suggested that Suisun Marsh supported one of the densest populations of river otters in California, and they are still abundant in the Marsh (authors' personal observations). River otter abundance may be related to present management of duck clubs and wildlife areas, although otters move freely among diverse habitats in the Marsh. The nontidal marshes, however, provide extensive, productive ditch and pond habitat for otter foraging. Prey consists primarily of crayfish, although waterfowl and fish are also taken. Consumption of waterfowl is proportional to waterfowl density; increased fish consumption was noted as migrating birds left the Marsh (Grenfell 1974).

American Beaver. Beaver populations have at least partially recovered from overexploitation (see chapter 2) following the ban on trapping. They can now be found foraging and breeding in slough habitats where tules and cattails are pres-

FIGURE 6.3. Beaver lodge in Suisun Marsh (photo by A. Manfree).

ent (Olofson 2000). In the Marsh they appear to feed largely on tule and cattail roots and build large conspicuous lodges from these materials (figure 6.3). Their role in Marsh ecology is not well understood, but elsewhere they are known as ecosystem engineers for their effects on hydrology, vegetation, and dikes.

Amphibians and Reptiles

Suisun Marsh and the nearby upland areas are presumably home to a number of reptiles and amphibians, but their populations there are poorly documented. Here, we discuss only Pacific pond turtle and frogs.

Pacific Pond Turtle. The Pacific pond turtle is a Species of Special Concern in California. Its use of estuarine habitats is poorly understood because it is mostly a freshwater species. In Suisun Marsh, these turtles use the upper reaches of tidal sloughs and managed wetlands and lay eggs in upland areas (USBR et al. 2011), including dikes. As the marsh becomes saltier, its populations are likely to be pushed more and more to the upper, fresher margins and to decline as a consequence. Populations have declined across the state; this decline is linked to a wide variety of factors, mostly related to habitat loss and alien species (Olofson 2000).

Frogs. The California red-legged frog is a federally listed threatened species found in vernal pool, riparian, and upland habitats in its lowland range throughout California. Small numbers have been found in the western portion of Suisun Marsh (Olofson 2000). Most of the Marsh, however, is not suitable habitat for this species (USBR et al. 2011). At present, alien bullfrogs are present in many of the freshwater areas of the Marsh, where they presumably exclude the native red-legged frog. The most common amphibian in the Marsh, however, is the native Pacific chorus frog, which thrives in a wide variety of habitats as long as freshwater pools are present in spring for breeding.

Invertebrates

We have largely ignored terrestrial invertebrates in the Marsh, in part because they are even more poorly known than nonwaterfowl terrestrial vertebrates. An exception is the butterfly (Lepidoptera) fauna (box 6.2). We are assuming, perhaps erroneously, that management that favors native terrestrial vertebrates will also favor the native invertebrates.

THE FUTURE OF TERRESTRIAL HABITATS AND SPECIES

Management of terrestrial, especially wetland, habitat in Suisun Marsh has been driven largely by the needs of ducks and geese that can be hunted (see chapter 5). This emphasis will continue as long as duck clubs and duck hunting exist in the Marsh, but other terrestrial vertebrates increasingly influence management efforts. While listed species, such as salt marsh harvest mouse and California clapper rail, have mandated protection, charismatic species, such as tule elk, river otters, white pelicans, and shorebirds, are increasingly important for managers. This is especially the case on public land, where hunting is eclipsed by wildlife viewing and other kinds of recreation as important human activities. Thus, it is important to consider the effects of large-scale change (as discussed in chapters 3, 8, and 9) on terrestrial vertebrates in a Marsh that will become increasingly brackish and tidal. Vertebrate population responses to these changes will greatly depend on management strategies put in place now.

Sea-level Rise

In the San Francisco Estuary, sea-level rise in the next 100 years will likely convert much of the intertidal habitat to subtidal habitat (chapter 3). Large amounts of marsh and mudflat habitat will likely become submerged, and protective dikes at

BOX 6.2. SUISUN BUTTERFLIES:
YESTERDAY, TODAY, TOMORROW?

Arthur M. Shapiro

My Suisun Marsh field site, along Peytonia Slough adjacent to the Suisun City Marina, is the oldest of 10 California sites where I monitor butterfly faunas (http://butterfly.ucdavis.edu). I started going there in 1972, as a stop on a transect across California following the I-80 corridor, a transect that would incorporate all the state's major climate zones, including the coastal fog belt. But I realized all too quickly that predicting good butterfly weather at the coast from a base in Davis was a hopeless enterprise: the fog was far too capricious for that. Suisun Marsh, then, was a reasonable substitute: cooled by the sea breeze through Carquinez Straits but almost never foggy in summer, and easily accessible.

It soon became apparent that the Marsh was a special place for butterflies, even if its fauna was fairly limited (Shapiro 1974). It had distinctive local races of two species, the field crescent (*Phyciodes campestris*) and the purplish copper (*Lycaena helloides*), and a thriving population of the highly colonial Yuma skipper (*Ochlodes yuma*). It had a population of America's smallest butterfly, the pygmy blue (*Brephidium exile*), that sometimes numbered in the tens of thousands in autumn. It was a stopover for migrating monarchs (*Danaus plexippus*) heading for the coast and for painted ladies (*Vanessa cardui*) heading for the desert to overwinter. By 2012, sixty species had been recorded in my study site. According to many temperate-zone standards, that's not bad at all! Only 58 species are known from the entire United Kingdom. By California standards, however, it's pretty ho-hum; at least 118 species have been recorded at Donner Summit, for example. Of the 60, a minimum of 19 are not regular breeding residents, and a few of these are long-distance "vagrants" from Southern California (Shapiro and Manolis 2007).

But in many ways the greatest value of Suisun Marsh for my research is the window it gives into the ecology of freshwater marsh butterflies before Euro-American colonization. Many of California's plant communities, particularly wetlands, have been converted to agriculture and urban centers, or degraded through human activities, including alien species introductions. Thus, finding reasonably undisturbed native wetland vegetation to study native butterfly life histories is no easy task. A major problem in understanding the California butterfly fauna is the poor fit of the Central Valley fauna to its climate and native vegetation. Except for riparian tree feeders, which are largely multiple-brooded, most lowland California butterflies have life histories adapted to the dry-summer Mediterranean climate (i.e., they have 1 or 2 broods in spring and then go dormant). Not so the Central Valley fauna, which is almost entirely composed of species that breed continuously from late winter to after Halloween. Today their larval hosts are weed species dependent on irrigation. So what did the larvae eat before Europeans arrived with their weeds? The answer came from careful observation of these species in remnants of low-elevation wetlands, particularly Suisun Marsh. It is here, for example, that I could find the anise swallowtail (*Papilio zelicaon*) still breeding on native Apiaceae, including water parsley (*Oenanthe sarmentosa*) and water hemlock (*Cicuta* spp.), even though probably more than 90% of its breeding today is on an alien perennial weed, sweet fennel (*Foeniculum vulgare*). I infer that a substantial portion of our Central Valley butterfly fauna was recruited from tule marshes, which stayed green in summer. The butterflies spread out over the Valley when irrigation and edible weeds allowed this to happen (Graves and Shapiro 2003; Shapiro 2009). A very spectacular example is the Mylitta crescent (*Phyciodes mylitta*), which feeds on the now endangered Suisun thistle (*Cirsium hydrophilum* var. *hydrophilum*) in the Marsh and also lives on weedy Old World annual thistles in every patch of disturbed ground in the region.

Only two locally breeding species, the sooty wing (*Pholisora catullus*) and the large marble (*Euchloe ausonides*), have gone extinct in the period of study at Suisun Marsh. Both have declined throughout north-central California, and the reasons for this are not understood. The field crescent also has declined regionally. It disappeared from my Suisun site for 5 years (15 generations), and I assumed that it was locally extinct. But this butterfly reappeared for 2 years— and then disappeared again. Presumably it is out there breeding, somewhere. A number of other butterfly species, including the great copper (*L. xanthoides*) and the least checkered skipper (*Pyrgus scriptura*), are in decline regionally, including in Suisun Marsh, for unknown reasons.

The Suisun Marsh butterfly fauna is overwhelmingly a freshwater marsh fauna. Increasing salinity in the Marsh will lead to loss of the resources on which this fauna depends. Pickleweed (*Sarcocornia pacifica*) marsh has only one butterfly, the pygmy blue, and ultimately it may be the sole survivor of this intriguing fauna and a beneficiary of climate change!

the urban edges will exacerbate the problem. In addition to direct loss of habitat, tidal marsh vertebrates are especially threatened because human actions reduce the natural functioning of marshes, including the ability to cope with sea-level changes of great magnitude (Erwin et al. 2006; Takekawa et al. 2006). Sea-level rise is also associated with increasing storm surges in California, causing greater erosion of shorelines and mudflats on which many vertebrates depend. Habitat for some species that depend on high marsh, such as black rail and Suisun ornate shrew, will have to be developed at higher elevations, unless subsidence reversal is successful on a large scale and safe "island" refuges from high tides and storms are developed. Unfortunately, areas for expansion of upper marsh habitat and transitional upland habitat are in short supply, although the Suisun region offers more potential than other more urbanized portions of the Estuary.

Earthquakes

Several faults in the region have the potential to significantly impact Suisun Marsh (chapters 3 and 8). Earthquakes of even moderate magnitude can cause sudden levee breaks that, combined with sea-level rise, could cause loss of habitat as the Marsh area is flooded. A rapid increase in flooded habitat in spring would submerge nests of marsh birds (e.g., clapper and black rails, Suisun song sparrows) and drown nestlings. Sudden water influx may also affect small mammals (e.g., salt marsh harvest mouse, Suisun shrew), amphibians, and reptiles that are limited in their ability to move and disperse rapidly. Additionally, salinity changes from an incursion of salt water, if the breaches occur during late summer or fall, can alter plant community composition as well as decrease freshwater availability. Some bird species in the Marsh are dependent on the freshwater ponds and ditches present in duck clubs and wildlife areas, so developing alternative freshwater areas (e.g., upper Denverton Slough) may be important to reduce the impacts of catastrophic dike failures.

Subsidence

Subsidence of managed lands on the Marsh plain is increasingly a problem; it can exacerbate the effects of sea-level rise, storm flooding, and decline in sediment deposition. Subsidence is largely created by waterfowl management strategies that cause oxidation of peaty sediments and reverse accretion of organic sediments through growth of marsh plants (chapter 3). Subsidence affects marsh vertebrates indirectly, because it interacts with other factors and makes restoration to previous conditions more difficult, and directly because it contributes to loss of productive mudflat edge habitat, important for shorebirds.

Climate Change

Climate change in the Marsh will likely include warmer temperatures, a more saline environment for longer periods as a result of sea-level rise, more extensive droughts, and periodic major floods. In general, global warming will also cause increased frequency and severity of tidal storm surges, leading to increased shoreline erosion and habitat loss (chapters 3 and 8). Warmer temperatures and changes in rainfall patterns can change breeding and migration times of birds and mammals, which may affect Marsh management. The increase in frequency and severity of large storm events have potential to undo higher and stronger dikes built to protect interior marsh habitats for duck production; such storms are also likely to create a more variable pattern of salinity in the Marsh (chapters 3 and 9).

SUMMARY AND CONCLUSIONS

The ability of Suisun Marsh to support its present assemblage of terrestrial vertebrates and invertebrates is almost certainly going to change dramatically, from forces originating both inside and outside the Marsh. At present Suisun Marsh is a highly altered landscape that provides habitat for a wide range of terrestrial vertebrates, native and non-native. It is especially important for those vertebrates that require wetlands, wildlife habitats that are in increasingly short supply in the San Francisco Estuary. The importance of the Marsh to waterfowl has long been acknowledged, but its importance to other terrestrial vertebrates, especially birds and mammals, is only beginning to be fully appreciated. Likely important drivers of future management decisions include protecting declining endemic species such as salt marsh harvest mouse, Suisun ornate shrew, and Suisun song sparrow as well as listed species such as California clapper rail, California black rail, and California least tern. As the Marsh grows in importance for natural history tourism, charismatic animals such as tule elk, river otters, white pelicans, and herons and egrets will rise in importance, so managing the Marsh as a mixture of tidal and nontidal habitats will be important. It is not possible to recover the huge amounts of marsh and wetland habitat lost in the Estuary and the state, but Suisun Marsh is clearly a key component in any management scheme to protect biodiversity, as a habitat island, as a node in a complex of reserves, and as a destination for migratory animals, especially birds. It will be increasingly important as a regional center of terrestrial biodiversity that connects to other natural areas in the region, especially to the north. Management of Suisun Marsh in ways that take into account projected large-scale changes in habitat both inside and outside of its boundaries should help curb future species losses.

REFERENCES

Ackerman, J. T. 2002. Of mice and mallards: positive indirect effects of coexisting prey on waterfowl nest success. Oikos 99: 469–480.

Albertson, J. D. 1995. Ecology of the California clapper rail in south San Francisco Bay. MS thesis, San Francisco State University, San Francisco, California.

Arnold, A. 1996. Suisun Marsh History: Hunting and Saving a Wetland. Marina, California: Monterey Pacific.

Beedy, E. C. 2008. Tricolored Blackbird *Agelaius tricolor*. Pages 437–443 *in* California Bird Species of Special Concern: A Ranked Assessment of Species, Subspecies, and Distinct Populations of Birds of Immediate Conservation Concern in California (W. D. Shuford and T. Gardali, eds.). Studies of Western Birds No. 1. Western Field Ornithologists, Camarillo, California, and California Department of Fish and Game, Sacramento.

Bias, M., and M. Morrison. 2006. Habitat selection of the salt marsh harvest mouse and sympatric rodent species. Journal of Wildlife Management 70:732–742.

California Department of Fish and Game. 2004. Suisun Marsh vegetation mapping change detection: A report to the California Department of Water Resources. California Department of Fish and Game, Sacramento.

California Department of Water Resources. 2011. Wildlife of the Suisun Marsh, Golden Eagle. http://www.water.ca.gov/suisun/dataReports/docs/SEW/geagle.cfm.

Choquenot, D., J. McIlroy, and T. Korn. 1996. Managing vertebrate pests: feral pigs. Bureau of Resource Sciences. Canberra: Australian Government Publishing Service.

Daehler, C. C., and D. R. Strong. 1997. Hybridization between introduced smooth cordgrass (*Spartina alternifolia;* Poaceae) and native California cordgrass (*S. foliosa*) in San Francisco Bay, California. American Journal of Botany 84: 607–611.

Del Giudice-Tuttle, E., K. Johston, and I. Medel. 2011. Baseline Assessment Program: 2009–2010 Report. Ballona Wetlands Ecological Reserve. Santa Monica Bay Restoration Commission. California State Coastal Conservancy, Los Angeles.

Erwin, R. M., G. M. Sanders, D. J. Prosser, and D. R. Cahoon. 2006. High tides and rising seas: potential effects on estuarine waterbirds. Studies in Avian Biology 32:214–228.

Evens, J. [G], and N. Nur. 2002. California black rails in the San Francisco Bay region: spatial and temporal variation in distribution and abundance. Bird Populations 6: 1–12.

Evens, J. [G.], and G. Page. 1986. Predation on black rails during high tides in salt marshes. Condor 88: 107–109.

Evens, J. G., G. W. Page, S. A. Layon, and R. W. Stallcup. 1991. Distribution, relative abundance, and status of the California black rail in western North America. Condor 93: 952–966.

Evens, J. G., G. W. Page, L. E. Stenzel, R. W. Stallcup, and R. P. Henderson. 1989. Distribution and relative abundance of the California black rail (*Laterallus jamaicensis coturniculus*) in tidal marshes of the San Francisco estuary. Report to the California Department of Fish and Game. Point Reyes Bird Observatory, Stinson Beach, California.

Foerster, K. S., J. E. Takekawa, and J. D. Albertson. 1990. Breeding density, nesting habitat, and predators of the California clapper rail. Report SFBNWR-116400–90–1. San Francisco Bay National Wildlife Refuge, Fremont, Caifornia.

Foin, T. C., E. J. Garcia, R. E. Gill, S. D. Culberson, and J. N. Collins. 1997. Recovery strategies for the California clapper rail (*Rallus longirostris obsoletus*) in the heavily-urbanized San Francisco estuary ecosystem. Landscape and Urban Planning 38: 229–243.

Gardali, T., and J. Evens. 2008. San Francisco Common Yellowthroat *Geothlypis trichas sinuosa*. Pages 346–350 *in* California Bird Species of Special Concern: A Ranked Assessment of Species, Subspecies, and Distinct Populations of Birds of Immediate Conservation Concern in California (W. D. Shuford and T. Gardali, eds.). Studies of Western Birds No. 1. Western Field Ornithologists, Camarillo, California, and California Department of Fish and Game, Sacramento.

Gervais, J. A., D. K. Rosenberg, and L. A. Comrack. 2008. Burrowing Owl *Athene cunicularia*. Pages 218–226 *in* California Bird Species of Special Concern: A Ranked Assessment of Species, Subspecies, and Distinct Populations of Birds of Immediate Conservation Concern in California (W. D. Shuford and T. Gardali, eds.). Studies of Western Birds No. 1. Western Field Ornithologists, Camarillo, California, and California Department of Fish and Game, Sacramento.

Gill, J. R. 1979. Status and distribution of the California clapper rail (*Rallus longirostris obsoletus*). California Fish and Game 65: 36–49.

Graves, S. D., and A. M. Shapiro. 2003. Exotics as host plants of the California butterfly fauna. Biological Conservation 110: 413–433.

Greenberg, R., C. Elphick, J. C. Nordby, C. Gjerdrum, H. Spautz, G. Shriver, B. Schmeling, B. Olson, P. Marra, N. Nur, and W. Winter. 2006a. Flooding and predation: trade-offs in the nesting ecology of tidal-marsh sparrows. Studies in Avian Biology 32: 96–109.

Greenberg, R., J. Maldonado, S. Droege, and M. V. McDonald. 2006b. Tidal marshes: a global perspective on the evolution and conservation of their terrestrial vertebrates. BioScience 56: 675—685.

Grenfell, W. E., Jr. 1974. Food habits of the river otter in Suisun Marsh, Central California. MS thesis, California State University, Sacramento.

Grinnell, J., H. C. Bryant, and T. I. Storer. 1918. The game birds of California. Contributions from the Museum of Vertebrate Zoology. Berkeley: University of California Press.

Grinnell, J., J. S. Dixon, and J. M. Linsdale. 1937. Fur-bearing mammals of California: their natural history, systematic status, and relations to man. Contributions from the Museum of Vertebrate Zoology. Berkeley: University of California Press.

Harvey, T. E. 1988. Breeding biology of the California Clapper Rail in south San Francisco Bay. Transactions of the Western Section of the Wildlife Society 24: 98–104.

Harvey, T. E., K. J. Miller, R. L. Hothem, M. J. Rauzon, G. W. Page, and R. A. Keck. 1992. Status and trends report on wildlife of the San Francisco Estuary. USDI Fish and Wildlife Service report for the San Francisco Estuary Project. San Francisco: U.S. Environmental Protection Agency and San Francisco Bay Regional Water Quality Agency.

Hays, W. S. 1990. Population ecology of ornate shrews, *Sorex ornatus*. MA thesis, University of California, Berkeley.

Houghteling, J. C. 1976. Suisun Marsh Protection Plan. San Francisco Bay Conservation and Development Commission, San Francisco.

Humple, D. 2008. Loggerhead shrike *Lanius ludovicianus* (mainland populations). Pages 271–277 *in* California Bird Species of Special Concern: A Ranked Assessment of Species, Subspecies, and Distinct Populations of Birds of Immediate Conservation Concern in California (W. D. Shuford and T. Gardali, eds.). Studies of Western Birds No. 1. Western Field Ornithologists, Camarillo, California, and California Department of Fish and Game, Sacramento.

Kelly, J. P., K. Etienne, C. Strong, M. McCaustland, and M. L. Parkes. 2007. Status, trends, and implications for the conservation of heron and egret nesting colonies in the San Francisco Bay Area. Waterbirds 30: 455–478.

Kelly, J. P., H. M. Pratt, and P. L. Greene. 1993. The distribution, reproductive success, and habitat characteristics of heron and egret breeding colonies in the San Francisco Bay area. Colonial Waterbirds 16: 18–27.

Larsen, C. J. 1989. A status review of the Suisun song sparrow (*Melospiza melodia maxillaris*) in California. Candidate Species Status Report 89–6. California Department of Fish and Game, Wildlife Management Division, Sacramento.

Lightfoot, K., and O. Parrish. 2009. California Indians and Their Environment: An Introduction. Berkeley: University of California Press.

Liu, L., P. Abbaspour, M. Herzog, N. Nur, and N. Warnock. 2007. San Francisco Bay Tidal Marsh Project annual report 2006: Distribution, abundance, and reproductive success of tidal marsh birds. Petaluma, California: PRBO Conservation Science.

Mall, R. E. 1969. Soil-water-plant relationships of waterfowl food plants in the Suisun Marsh of California. Wildlife Bulletin No. 1. California Department of Fish and Game, Sacramento.

Marshall, J. T., and K. G. Dedrick. 1994. Endemic song sparrows and yellowthroats of San Francisco Bay. Studies in Avian Biology 15: 316–327.

Marvin-DiPasquale, M. C., J. L. Agee, R. M. Bouse, and B. E. Jaffe. 2003. Microbial cycling of mercury in contaminated pelagic and wetland sediments of San Pablo Bay, California. Environmental Geology 43: 260–267.

Newman, J. R. 1970. Energy flow of a secondary consumer (*Sorex sinuosus*) in a salt marsh community. PhD dissertation, University of California, Davis.

Novak, J. M., K. F. Gaines, J. C. Cumbee, G. Mills, A. Rodriguez-Navarro, and C. S. Romanek 2005. The clapper rail as an indicator species of estuarine-marsh health. Studies in Avian Biology 32: 270–281.

Nur, N., S. Zack, J. Evens, and T. Gardali. 1997. Tidal marsh birds of the San Francisco Bay Region: status distribution, and conservation of five Category 2 taxa. Draft final report to the United States Geological Survey, Biological Resources Division. Point Reyes Bird Observatory, Stinson Beach, California.

Olofson, P. R., ed. 2000. Baylands Ecosystem Species and Community Profiles: Life Histories and Environmental Requirements of Key Plants, Fish and Wildlife. Prepared by the San Francisco Bay Area Wetlands Ecosystem Goals Project. San Francisco Bay Regional Water Quality Control Board, Oakland, California.

Parris, A., P. Bromirski, V. Burkett, D. Cayan, M. Culver, J. Hall, R. Horton, K. Knuuti, R. Moss, J. Obeyserka, A. Sallenger, and J. Weiss. 2012. Global sea level rise scenarios for the US National Climate Assessment. NOAA Technical Memo OAR CPO-1.

Resh, V. H. 2001. Mosquito control and habitat modification: case history studies of San Francisco Bay wetlands. Pages 413–428 *in* Biomonitoring and Management of North

American Wetlands (R. B. Rader, D. P. Batzer, and S. Wissinger, eds.). New York: John Wiley and Sons.

Roberson, D. 2008. Short-eared owl *Asio flammeus*. Pages 242–248 *in* California Bird Species of Special Concern: A Ranked Assessment of Species, Subspecies, and Distinct Populations of Birds of Immediate Conservation Concern in California (W. D. Shuford and T. Gardali, eds.). Studies of Western Birds No. 1. Western Field Ornithologists, Camarillo, California, and California Department of Fish and Game, Sacramento.

San Francisco Estuary Institute. 2003. Pulse of the Estuary: Monitoring and Managing Contamination in the San Francisco Estuary. Oakland: San Francisco Estuary Institute.

Schwarzbach, S., and T. Adelsbach. 2002. Assessment of ecological and human health impacts of mercury in the bay-Delta watershed, Subtask 3B: field assessment of avian mercury exposure in the bay-delta ecosystem. Final Report to the CALFED Bay–Delta Mercury Project. CALFED Bay–Delta Authority, Sacramento.

Schwarzbach, S. E., J. D. Albertson, and C. M. Thomas. 2006. Effects of predation, flooding, and contamination on reproductive success of California clapper rails (*Rallus longirostris obsoletus*) in San Francisco Bay. Auk 123: 1–16.

Scollon, D. B. 1993. Spatial analysis of the tidal marsh habitat of the Suisun Song Sparrow. MA thesis, San Francisco State University, San Francisco, California.

Shapiro, A. M. 1974. Butterflies of the Suisun Marsh. Journal of Research on the Lepidoptera 13: 191–206.

Shapiro, A. M. 2009. Revisiting the pre-European butterfly fauna of the Sacramento Valley, California. Journal of Research on the Lepidoptera 41: 31–39.

Shapiro, A. M., and T. D. Manolis. 2007. Field Guide to Butterflies of the San Francisco Bay and Sacramento Valley Regions. Berkeley: University of California Press.

Shellhammer, H. S. 1989. Salt marsh harvest mice, urban development, and rising sea levels. Conservation Biology 3: 59–65.

Shuford, W. D., and T. Gardali, eds. 2008. California Bird Species of Special Concern: A Ranked Assessment of Species, Subspecies, and Distinct Populations of Birds of Immediate Conservation Concern in California. Studies of Western Birds No. 1. Western Field Ornithologists, Camarillo, California, and California Department of Fish and Game, Sacramento.

Skinner, J. E. 1962. An historical review of the fish and wildlife resources of the San Francisco Bay Area. Water Projects Branch Report 1. California Department of Fish and Game, Sacramento.

Spautz, H., L. Liu, S. Estrella, and N. Nur. 2012. Population studies of tidal marsh breeding birds at Rush Ranch: a synthesis. San Francisco Estuary and Watershed Science 10(2).

Spautz, H., and N. Nur. 2008. Suisun song sparrow *Melospiza melodia maxillaris*. Pages 405–411 *in* California Bird Species of Special Concern: A Ranked Assessment of Species, Subspecies, and Distinct Populations of Birds of Immediate Conservation Concern in California (W. D. Shuford and T. Gardali, eds.). Studies of Western Birds No. 1. Western Field Ornithologists, Camarillo, California, and California Department of Fish and Game, Sacramento.

Spautz, H., N. Nur, D. Stralberg, and Y. Chan. 2006. Multiple-scale habitat relationships of tidal-marsh breeding birds in the San Francisco Bay estuary. Studies in Avian Biology 32: 247–269.

Sustaita, D., P. F. Quickert, L. Patterso, L. Barthman-Thompson, and S. Estrella. 2011. Salt marsh harvest mouse demography and habitat use in the Suisun Marsh, California. Journal of Wildlife Management 75: 1498–1507.

Takekawa, J. I. W., H. Spautz, N. Nur, J. L.Greenier, K. Malamud-Roam, J. C. Nordby, A. Cohen, F. Malamud-Roam, and S. E. Wainwright-De la Cruz. 2006. Environmental threats to tidal marsh vertebrates of the San Francisco Bay Estuary. Studies in Avian Biology 32: 176–197

U.S. Bureau of Reclamation, U.S. Fish and Wildlife Service, and California Department of Fish and Game. 2011. Suisun Marsh Habitat Management, Preservation, and Restoration Plan Final environmental impact statement/environmental impact report. U.S. Bureau of Reclamation, Sacramento.

U.S. Fish and Wildlife Service. 2009. Draft recovery plan for tidal marsh ecosystems of northern and central California. U.S. Department of Interior, Fish and Wildlife Service, Sacramento.

WESCO. 1986. A review of the population status of the Suisun shrew *Sorex ornatus sinuosus*. Final Report (FW 8502), USFWS. U.S. Department of Interior, Fish and Wildlife Service, Sacramento.

Wiener, J. G., C. C. Gilmour, and D. P. Krabbinhoft. 2003. Mercury strategy for the bay–delta ecosystem: a unifying framework for science, adaptive management, and ecological restoration. Draft Final Report to CALFED. CALFED Bay–Delta Authority, Sacramento.

Wiener, J. G., and D. J. Spry. 1996. Toxicological significance of mercury in freshwater fish. Pages 297–339 *in* Environmental Contaminants in Wildlife: Interpreting Tissue Concentrations (W. N. Beyer, G. H. Heinz, and A. W. Redmon- Norwood, eds.). Boca Raton, Florida: Lewis.

Wolfe, M. F., S. Schwarzbach, and R. A. Sulaiman. 1998. Effects of mercury on wildlife: a comprehensive review. Environmental Toxicology and Chemistry 17: 146–160.

7

Fishes and Aquatic Macroinvertebrates

Teejay A. O'Rear and Peter B. Moyle

Suisun Marsh contains some of the most important habitats for fishes and macroinvertebrates in the San Francisco Estuary (Moyle et al. 2012). More than 50 species of fish have been collected from the Marsh, 27 of which occur frequently enough to be part of at least seasonal fish assemblages, along with eight species of shrimp, clams, and other macroinvertebrates (scientific names of all fishes and macroinvertebrates mentioned in the text are provided in chapter appendix 7.1). This is partly due to its intermediate location in the Estuary where salinity, temperature, and water clarity are often within the optimum ranges of fishes such as striped bass, delta smelt, longfin smelt, and splittail. At the same time, these factors are variable enough to exclude less tolerant alien fishes, such as largemouth bass and bluegill, and to limit the distribution of harmful invertebrates such as overbite clam. Suisun Marsh also serves as a corridor for migratory fishes: adult white sturgeon, striped bass, Chinook salmon, delta smelt, and longfin smelt pass through its sloughs on their way upstream to spawn, and juveniles move through them to forage and rear.

This chapter examines the historical fish assemblages of Suisun Marsh and the major ecological factors that have structured them. It then turns to environmental changes likely to affect the fish assemblages, what the future community may look like, and what we can do to make the Marsh of the future a more benign environment for threatened and endangered fishes.

SUISUN MARSH AS HABITAT

Suisun Marsh is located in the middle reaches of the San Francisco Estuary, so its aquatic habitats support a largely estuarine fauna, tolerant of a wide range of conditions. Salinities vary substantially over a range of temporal scales, including daily (due to tidal action), monthly (due to Delta outflow), annual (due to the type of hydrological year), and decadal (due to long-term droughts). Diversity and abundance of fishes and macroinvertebrates are enhanced by complex habitats within the tidal sloughs. These sloughs are often sinuous, have convoluted edges, and are relatively shallow, providing a relatively greater amount of complex edge habitat than is found in most other areas of the Estuary, especially the Sacramento–San Joaquin Delta. In addition, some sloughs have connections to undiked marsh plains, which provide both additional habitat for fish and a potential source of nutrients and invertebrates for export into the sloughs.

Although the Marsh functions as important habitat for all life-history stages of fishes, the tidal sloughs are perhaps most valuable as nursery areas for juveniles. Shallow sloughs and vegetated dikes provide both a refuge from predation and invertebrate food organisms such as amphipods for fishes. The low water clarity also provides small fish with protection from visually feeding predators such as large striped bass, terns, and herons. The generally shallow, turbid environment, coupled with complex channel networks and high residence times, promotes high concentrations of zooplankters that serve as critical food for both smaller juvenile fishes and all life-history stages of planktivores such as delta smelt. Those same zooplankters also are eaten by opossum shrimp and larger grass shrimp, which, in turn, are fed upon by larger juvenile and adult fishes. It is also likely that the well-vegetated banks of some sloughs provide foraging habitat for juvenile Chinook salmon and steelhead

Although the sloughs are major fish habitat, large expanses of aquatic habitat are behind the dikes of duck clubs and in wildlife areas. While many of these ditches and ponds are seasonal, others provide permanent habitat for native species such as threespine stickleback and prickly sculpin, as well as for alien species such as common carp. Movement of fish between slough and interior habitats is limited; it occurs mainly through pipes and gated culverts when hunting areas are being flooded or drained. At times, large numbers of small fish, often sticklebacks, are flushed from the interior marshes, and predators such as striped bass eat these small fish while holding within the outflow plumes (authors' unpublished data). Unfortunately, water discharged from these marshy areas, when coupled with unfavorable conditions such as neap tides, weak winds, and warm water temperatures, can create low-oxygen or anoxic conditions in the sloughs. These black-water events, which usually take place in the fall, can cause fish kills and can render substantial portions of some sloughs inhospitable to most life (Siegel et al. 2011).

While the Marsh sloughs still provide good habitat for fishes and macroinvertebrates, the ecological functioning of that habitat has declined, concurrently

FIGURE 7.1. Staghorn sculpin (*Leptocottus armatus*; photo by A. Manfree).

with habitat changes that have occurred in the Delta. Phytoplankton abundances in the larger sloughs have been in long-term decline, as have numbers of opossum shrimp (Schroeter 2010). Likewise, the long-term decreases in abundances of delta and longfin smelt seen in other parts of the Estuary have been paralleled in the Marsh. However, the situation in Suisun Marsh does not seem to be quite as dire as that in the Delta—both striped bass and threadfin shad do not have long-term declining trends in the Marsh as they do in the Delta (O'Rear and Moyle 2010, 2011). Consequently, it seems that while the quality of the habitat in the Marsh has decreased over time, that drop has not been as substantial in some respects as that seen in the Delta.

The future importance of the Marsh's slough habitats to fishes, especially those valued by humans, is likely to change from what it is today. First, the dual climate-change effects of sea-level rise and warmer temperatures will make the water warmer and saltier than it is now. However, assuming that both droughts and floods will be more frequent (see chapter 3), the variability in salinity is likely to increase, although most of the time, average salinity will be higher. As a corollary, because many of the dikes in Suisun Marsh have very little freeboard, are constructed of easily erodible materials (e.g., peat), and surround land that is subsided, the amount of open-water habitat is likely to increase beyond that which is currently planned for. Additionally, the more saline environment of the Marsh may favor the expansion of the filter-feeding alien overbite clam, which could reduce the functioning of the Marsh as a producer of planktonic food. These environmental changes will likely lead to an *increase* in native species as a portion of the fish community; however, the native fishes that benefit from the altered conditions will probably not be species of most concern today, such as delta smelt and longfin smelt, but rather more marine benthic fishes such as staghorn sculpin, longjaw mudsucker, and starry flounder (see figure 7.1).

PLANKTON, MACROINVERTEBRATES, AND FISHES—
THE PAST 30 YEARS

Monitoring programs have given a reasonably clear, although admittedly incomplete, picture of changes in the Marsh's fishes and invertebrates in recent decades. One, the Interagency Ecological Program's Environmental Monitoring Program, has documented trends in phytoplankton and zooplankton, key components of the pelagic food web (authors' unpublished data). Second, the Suisun Marsh Fish Study (1979–present) conducted by the University of California, Davis, has systematically surveyed fishes, shrimps, and, to a lesser extent, other large invertebrates such as clams and jellyfish (figure 7.2). These two programs have, in combination with short-term studies, revealed shifts in both the composition and the energy-flow pathways of the aquatic communities.

Plankton

Consistent with trends seen in the Delta, phytoplankton abundance has exhibited a long-term decline in annual abundance in the larger sloughs of the Marsh (e.g., Suisun and Montezuma sloughs; A. Müller-Solger, personal communication, 2004). Concurrent with these declines has been a general increase in the abundance of the overbite clam (Schroeter 2010), which has been shown to be a prodigious filter-feeder on both phytoplankton and smaller life-history stages of some zooplankters. On a finer scale, phytoplankton biomass though the year is negatively correlated with abundance of overbite clams. As a result, regions that see less fresh water through the year and/or have greater hydrologic connectivity to Grizzly Bay also have the lowest abundances of phytoplankton. Regions deeper in the interior of the Marsh that exhibit less connectivity and experience fresher water throughout much of the year have the highest abundances of phytoplankton and fewer overbite clams (Schroeter 2010).

Nevertheless, it is likely that phytoplankton abundance has generally declined throughout all regions of the Marsh in the past 30 years. The evidence is as follows. First, opossum shrimp abundance has declined, mirroring the long-term decline seen in the Delta. Much of the decline in opossum shrimp appears to be the result of poor survival of younger life-history stages that feed predominantly on phytoplankton. Second, many fishes in the Marsh switched from feeding primarily on opossum shrimp to eating other food items, especially gammarid and corophiid amphipods, during the decline (Feyrer et al. 2003). Amphipods are primarily benthic and are most abundant in structurally complex environments such as tule-root masses, beds of aquatic plants, and areas rich in detrital debris. Third, seasonal declines in opossum shrimp in the channels of the Marsh occur concurrently with both a decline of fishes in the channels and an increase of these same fishes in near-shore areas, a pattern consistent with fishes moving inshore

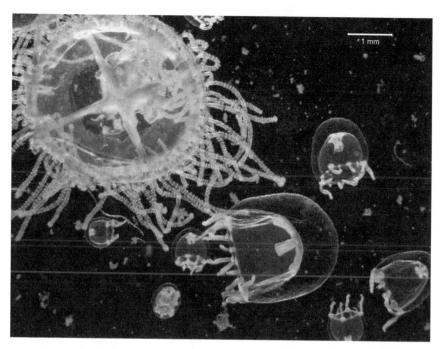

FIGURE 7.2. Alien hydrozoans from Suisun Marsh. A Black Sea jellyfish (*Maeotius marginata*) is on the upper left; the others are *Moerisia lyonsi* (photo by Alpa Wintzer).

to feed on amphipods (O'Rear and Moyle 2011). The net effect seems to be that more energy is now flowing through a benthic food web that has amphipods as a major node rather than opossum shrimp, with the corollary that phytoplankton is no longer the primary carbon source.

Macroinvertebrates

For larger invertebrates effectively captured by the trawl, long-term trends have been variable, with some species increasing and some decreasing. There has also been a succession of new invasions of alien species, with poorly understood effects on other species. Annual catch of native California bay shrimp generally has been on a downward slide, with some of the lowest annual numbers occurring just after the 2001 invasion and subsequent dramatic increase of Siberian prawn. The annual catch of the alien overbite clam, although highly variable, has been increasing over time. Black Sea jellyfish catches have also risen, although their numbers during 2010 and 2011 were small compared to their abundance in the mid-2000s. In general, numbers of California bay shrimp, Siberian prawn, and Black Sea jellyfish were low through the dry years of 2008–2010, while the highest-ever catches of overbite clam occurred in the same period. These patterns are fairly consistent with trends at lower levels of the pelagic food web.

The impact of recent macroinvertebrate invaders is hard to determine because so much else is going on simultaneously in the Marsh, such as the effects of upstream and in-marsh water management. Examples of the uncertainties include the following:

- The recent low catches of native bay shrimp appear to be due to competition with alien Siberian prawn whose catches have increased. However, differences in ecology of the two species, such as reproductive timing, diet, and physiology, suggest that abiotic parameters such as temperature and salinity are responsible for the trends of both shrimps in the Marsh. For example, Siberian prawns appear to feed primarily on benthic organisms while California bay shrimp prey on pelagic species (Siegfried 1982; Xu et al. 2008). Also, California bay shrimp are winter spawners in marine waters while Siberian prawn spawn in late spring and early summer in fresher regions (Hatfield 1985; Oh et al. 2002).
- The Marsh has been invaded in recent years by four predatory species of hydrozoans ("jellyfish," most conspicuously the Black Sea jellyfish [*Maeotias marginata*]; figure 7.2) (Wintzer et al. 2010, 2011a, 2011b). These species all prey on zooplankton and thus are potential competitors for plankton-feeding fishes, especially when the jellyfish are abundant during late summer and fall. Their peaks of seasonal abundance occur following peaks in abundance of larval fishes, so predation on fish larvae, although documented, does not seem to be affecting fish populations. However, the high genetic diversity of these hydrozoans and their ability to reproduce both asexually and sexually should allow them to adapt quickly to future conditions created by climate change, with greater negative effects than are currently documented.

Pelagic Fishes

In the Delta, pelagic fishes have declined, reflecting declines in phytoplankton and zooplankton. Four species in particular— striped bass, threadfin shad, longfin smelt, and delta smelt—have been treated together as part of the Pelagic Organism Decline (Sommer et al. 2007). While annual catches of native delta and longfin smelt in the Marsh follow the decline of these same species in the Delta, patterns in striped bass and threadfin shad catches have not exhibited the same downward slide. Some threadfin shad have always been present in the Marsh, regardless of type of water year; however, years with high Delta outflow in spring and early summer (e.g., 2006) often coincide with large increases in threadfin shad numbers. While these increases likely are due, in part, to recruitment of threadfin shad from upstream into the Marsh, they are also likely due to fresher

FIGURE 7.3. Sampling fish in Suisun Marsh with a seine net (photo by A. Manfree).

conditions favoring reproduction within the Marsh, as indicated by the higher abundance of larval, juvenile, and adult threadfin shad in smaller interior sloughs such as Denverton and First Mallard (O'Rear and Moyle 2011). These sloughs often provide more zooplankton food than larger marsh sloughs or the Delta, and this productivity, in combination with high turbidities that reduce predation, likely results in higher survival of juvenile fish.

Although the striped bass trawl catch declined through the 1980s and into the mid-1990s, no clear trends were evident from the mid-1990s through the 2000s. However, in recent years, a decline in the trawl catch often has been accompanied by an increase in the beach seine catch in dry years (figure 7.3), with the opposite occurring in wetter years (O'Rear and Moyle 2011). These changes often occur as opossum-shrimp abundance in the trawl also declines, which suggests that juvenile striped bass may be moving from channel to near-shore habitats to feed on amphipods and small fishes. Consequently, the apparent lack of a long-term decline in the striped bass catch in the Marsh may be due, in part, to food subsidies provided by vegetated inshore habitats during years lean in opossum shrimp.

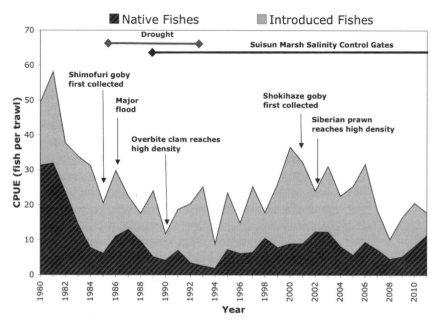

FIGURE 7.4. Trends in native and alien fishes in Suisun Marsh, 1980–2011, with years of major alien invasions and events (CPUE = catch per unit effort).

Alien Fishes

Long-term trends of alien fishes have been, like those of alien invertebrates, highly variable, although several patterns are apparent (figure 7.4). First, many of these species have had just a few years of high abundances bookended by low numbers in all other years, which in most cases has been due to interactions of flow, temperature, and salinity. Many of these species are freshwater fishes, such as black bullhead and black crappie, that recruit successfully only when high outflow and low salinity persist late in the year. However, two alien fishes that can tolerate somewhat higher salinities, white catfish and Mississippi silverside, have had long-term increasing trends in the Marsh.

Large populations of white catfish seem relatively innocuous for several reasons. White catfish have been present in the system since 1874, during which time they have coexisted with large populations of native fishes. Also, they feed primarily on amphipods, especially corophiids and a large gammarid, *Eogammarus confervicolus,* which do not seem to be limiting food resources. White catfish also rarely consume prey that may be limiting to threatened or endangered species (e.g., opossum shrimp, an important food of larger longfin smelt), they mainly feed on other fishes only when water management makes it profitable for them to do so, and they are a popular sport fish (O'Rear 2012). Mississippi silverside, on the other hand, present a greater threat because they often co-occur with delta and longfin smelt, their diet is similar to that of the smelts, they have been shown experimentally to feed on smelt larvae, and silversides have increased while the two smelt species have declined (Bennett and Moyle 1996). Finally, white catfish populations in the Marsh seem heavily dependent on fresh conditions during their reproductive period in late spring and early summer. Mississippi silverside populations, however, appear to be largely unaffected by the variability in abiotic conditions in the Marsh, with the exception of prolonged cold temperatures from autumn through spring that delay reproduction (O'Rear and Moyle 2011).

Native Fishes

Annual catch of native fishes has been more stable through time than that of alien fishes, with a decreasing trend from the early 1980s through the drought-stricken early 1990s and relatively stable catches since the wet years of the mid-1990s (Meng et al. 1994; Matern et al. 2002). By far the most abundant native fish in the Marsh, Sacramento splittail has a very clear positive relationship with high springtime flows that is reflected in large catches of young-of-year fish in the Marsh after spring in wet years (Matern et al. 2002). This relationship is driven by the success of spawning on floodplains upstream of the Delta (e.g., Yolo Bypass; Moyle et al. 2004). Young-of-year splittail appear to have relatively high survival, such that a strong year-class has the ability to mask poor reproductive years. Given high springtime flows that occur approximately every 4 years, splittail numbers—and hence, native fish numbers—stay relatively stable in the Marsh. The Sacramento sucker, which has many of the same life-history attributes as the splittail and is also an important component of the native fish community, has a similar pattern of abundance.

Considerable local variability in composition of fish assemblages has been caused by management practices of duck clubs within the Marsh. Although many of these managed wetlands are poor habitat for fishes in autumn because of low-oxygen or anoxic conditions, the ponds and ditches are likely very productive habitats once dissolved oxygen levels become acceptable during winter and spring. Further, they provide habitat required for spawning by two native fishes, the prickly sculpin and threespine stickleback. These two fishes reproduce in the managed wetlands during late winter and early spring. When the wetlands are drained in late spring, young-of-year sculpins and sticklebacks are discharged to the adjoining sloughs, such that local abundances of these species can be extremely high. Because movement of water in the managed wetlands is highly individualized and varies across space and time, the pulse of stickleback and prickly sculpin introductions into the Marsh's sloughs is also highly variable.

This discussion has highlighted the diverse factors structuring the fish and invertebrate assemblages of Suisun Marsh, especially salinity, temperature, water clarity, abundance of food, and Delta outflow. Although these factors are interlinked in myriad ways, sufficient data currently exist to provide an understanding of how these factors, and, consequently, the fish community of Suisun Marsh, will change in response to probable future conditions (see chapter 3).

THE FUTURE

The waters of Suisun Marsh will be generally warmer and saltier in the future, with greater variability in flows coming in from upstream, both seasonally and among years (chapters 3 and 9). These changes in physical and chemical conditions will cause changes to the assemblages of macroinvertebrates and of pelagic, benthic, and native fishes.

Macroinvertebrates

Higher salinities will likely increase both the extent and the abundance of overbite clam and hydrozoans in the Marsh and, consequently, reduce plankton densities. However, the clam does not appear to thrive in smaller sloughs that currently have appropriate salinities (e.g., upper Goodyear Slough), possibly because of high detritus loads that interfere with the clam's ability to filter feed, high abundances of splittail, a known predator of the overbite clam (Moyle et al. 2004), and reduced hydrologic connectivity to areas of clam production (e.g., lower Suisun Slough and Grizzly Bay). Although smaller interior sloughs may have higher salinities in the future, other aspects related to their geomorphology, such as soft bottoms, may still make them inhospitable to overbite clams. However, the larger sloughs (Montezuma, Suisun, and Nurse sloughs) will probably have higher clam and hydrozoan densities and thus lower plankton abundances. Nevertheless, frequent freshwater floods could keep clam populations at a low level, which would likely increase the Marsh's ability to support pelagic fishes that depend on plankton. Other macroinvertebrates will increase or decrease, depending on the species; given the history of invasions of San Francisco Estuary, it is quite likely that one or more new species will invade, with negative consequences.

Pelagic Species

Probably most important to future managers is how well the Marsh is likely to support at-risk pelagic species, especially delta and longfin smelt. Although delta smelt can persist in salinities up to about 18 ppt, they are nearly always most abundant around the 2-ppt isohaline and are rarely found in water with a salinity higher than 6 ppt (Feyrer et al. 2007b). Consequently, in most future years, the Marsh generally will be too salty to be much utilized by delta smelt. Given their greater tolerance of higher salinities, longfin smelt will likely derive more benefit from the Marsh of the future than delta smelt, but if overbite clams and jellyfish increase greatly in abundance, pelagic food supplies may become even more limiting for smelt. This may be partially mitigated by an increase in open-water habitat with patches of productivity, but the portion of the year when longfin smelt will be able to utilize the Marsh will likely decrease because higher water temperatures will restrict them to the cooler seasons. The importance of the Marsh as rearing habitat for both species could be increased in years with sustained high wintertime flows that freshen the Marsh significantly and, hence, reduce the population size of overbite clam.

Two other common pelagic species, threadfin shad and striped bass, are likely to be less abundant, at least seasonally, in the future. The frequency of recruitment of threadfin shad both within the Marsh and from upstream would no doubt decline, given the species' spawning requirement of fresh water. Juvenile striped bass are, like delta smelt, generally most abundant at the 2-ppt isohaline, and so, like delta smelt, should generally be more abundant upstream of the Marsh in most future years. Lower pelagic food production in the Marsh could be detrimental to very early life-history stages of striped bass that are dependent on copepods; however, larger juveniles would probably benefit from the greater amount of edge habitat and thus greater amphipod abundance.

Use of Suisun Marsh by adult striped bass is currently governed by temperature, with adults entering the Marsh once temperatures stay consistently below about 16°C in autumn and leaving to go upstream to spawn once temperatures exceed about the same value in spring. Thus, as for longfin smelt, the weeks of the year when adult striped bass will be in the Marsh will be more restricted in the future. As a result, striped bass will probably still be abundant in the Marsh, although the population will likely contain more large juveniles and fewer adults and small juveniles than it does currently.

If estuarine pelagic species decline, they will likely be replaced, in part, by more marine species, particularly Pacific herring, northern anchovy, and, to a lesser extent, other saltwater smelt species (e.g., surf smelt). A more saline environment and additional acreage of open-water habitat would be favorable for these species, but excessively warm water temperatures would restrict their use to the cooler seasons. Conversely, the higher salinities should pose no problem for the semipelagic Mississippi silverside, and warmer water temperatures would likely increase their numbers by lengthening their spawning period. A larger population of Mississippi silverside could put additional pressure on zooplankton populations, increasing the probability of competition among pelagic species. Thus, while native pelagic fishes would still be a part of the fish assemblage, their numbers would probably be lower because of increased competition with Mississippi silversides, clams, jellyfish, and each other for a reduced food supply.

Benthic Fishes

Many native benthic fishes will likely increase in the Marsh of the future. This prediction is based on several factors. First, upstream movement of saltier water in the Estuary would move appropriate spawning habitat of marine fishes, many of which are flatfishes, closer to the Marsh than is the case now. Second, several euryhaline marine species, such as staghorn sculpin, longjaw mudsucker, and starry flounder, are able to use benthic zones of open-water shallow habitats. Third, while the larger sloughs will likely have more overbite clams, they will probably still contain ample populations of other benthic invertebrates to support benthic-feeding fishes. Thus, given sufficiently high salinities, the channels should support marine fishes, such as white croaker and plainfin midshipman, that are common in channel habitats in other areas of the Estuary (e.g., South Bay). Finally, some of the more physiologically tolerant native fishes presently in the Marsh will likely remain common, including splittail, prickly sculpin, tule perch, and threespine stickleback. In particular, the abundance of splittail would depend on floodplains being inundated on a regular basis for spawning success to be high enough to maintain a large population (Moyle et al. 2004). Studies in Petaluma Marsh, which is generally more saline, indicate that splittail can live in surprisingly saline conditions (Feyrer et al. 2007a). In fact, the additional shallow habitat and increased edge perimeter in Suisun Marsh could contribute to greater foraging success, increasing splittail survival and growth.

Alien Species

Overall, alien species are expected to decline as a portion of the fish community in the Marsh, with the consequence that native fishes will become more dominant, represented especially by marine flatfishes, sculpins, and gobies. However, some small alien fishes (Mississippi silverside, rainwater killifish, and western mosquitofish) and annual benthic fishes (e.g., shimofuri goby) that are currently abundant will likely be even more prevalent in a future Marsh with warmer water temperatures and a greater expanse of shallow-water habitat, while catfishes and sunfishes will likely disappear because of high salinity. The alien fishes expected to increase have shorter life spans than either the catfishes or the sunfishes, so annual abundances of fishes may be even more variable than they are now.

In sum, the fish community of Suisun Marsh, reflecting abiotic conditions, will be considerably different in the future, although some of the more physiologically tolerant members will remain the same (see appendix 7.2). The Marsh will generally be too saline in most years for freshwater fishes such as catfishes and sunfishes while being more favorable to more marine benthic species that would also benefit from a greater area of open water. Similarly, a replacement of pelagic fishes will likely take place, with Pacific herring, northern anchovy, and saltwater smelt species replacing delta smelt, threadfin shad, and, to a lesser extent, longfin smelt and very young striped bass. However, the potentially increased abundance and extent of overbite clam coupled with a larger population of Mississippi silverside, despite a greater area of open-water habitat, may result in lower densities of native pelagic fishes due to competition for zooplankton. Nevertheless, years with very high flows that freshen the Marsh could be critical for maintaining habitat for at-risk species, highlighting the fact that the Marsh will remain important to efforts directed at conserving threatened and endangered fishes, as well as harvested species.

WHAT DO THE FISH NEED?

A future Suisun Marsh that favors native estuarine fishes should have the following interacting attributes:

1. High production of zooplankton
2. High abundance of benthic crustaceans, especially amphipods
3. A diversity of juvenile native fishes and striped bass that exhibit high growth and survival rates
4. Seasonal populations of delta smelt, longfin smelt, and juvenile Chinook salmon
5. Low densities of alien clams (overbite, Asian)
6. High variability of salinity in space and time
7. Levels of dissolved oxygen that never drop below 5 mg/L
8. Temperatures in at least some areas that stay below 22–24°C in summer
9. Tidal currents to move water and sediment
10. Freshwater inflows, year round, in most (but not all) years
11. A gradient of connectivity between interior marshes and tidal sloughs
12. A diversity of complex habitats along a salinity gradient
13. Biotic connections to habitats both upstream and downstream of the Marsh

A Marsh with these characteristics would support large populations of fish, be an important nursery area for fish and invertebrates, and be able to adapt to sea-level rise in the next hundred years. Some of these desired features will develop independently as sea level rises and dikes fail, creating more connections to interior marshes, ideally with dendritic tidal drainage networks. But others are likely to happen only with active management of flows, habitat, species, and vegetation (see chapter 9 for possibilities). It is likely that as large-scale changes take place throughout the entire San Francisco Estuary, Suisun Marsh will assume increasing importance as a refuge for the Estuary's more euryhaline aquatic biota. The more the future Marsh assumes the above attributes, the better it is likely to function as habitat for native fishes and alien estuarine fishes, such as striped bass and American shad, that support fisheries.

SUMMARY AND CONCLUSIONS

Suisun Marsh contains some of the most important habitats for fishes and aquatic invertebrates in the San Francisco Estuary. This is partly because of its complex slough habitats and its intermediate location in the Estuary where salinity, temperature, and water clarity are often within the optimum ranges of fishes such as striped bass, delta smelt, longfin smelt, and splittail. The diked nontidal marshes also provide habitat for some fishes, but connections depend on drainage practices. Despite favorable habitats, many invertebrate and fish species in the Marsh, especially native ones, are in decline, although alien shrimp, clams, and jellyfish are at least seasonally very abundant. The annual catch of native fishes has been more stable for the past 30 years than that of alien fishes, with a decreasing trend from the early 1980s through the drought of the late 1980s and early 1990s, and relatively stable catches since the wet years of the mid-1990s. The most abundant native fish in the Marsh, Sacramento splittail, has a positive relationship of abundance with high springtime flows. Some native fishes (e.g., threespine stickleback, prickly sculpin) are most abundant in the channels of the Marsh bordered by diked wetlands. The future fish community, in response to habitat changes wrought by sea-level rise and climate change, will retain some of the more environmentally tolerant members such as splittail. But the Marsh will generally be too saline in most years for freshwater fishes while being more favorable to more marine benthic species that would also benefit from a greater area of open water. Similarly, a replacement of pelagic fishes will likely take place, with Pacific herring, northern anchovy, and perhaps saltwater smelt species replacing delta smelt, threadfin shad, and, to a lesser extent, longfin smelt and very young striped bass. No matter what happens as a consequence of environmental change, Suisun Marsh will remain one of the most important regions in the Estuary for native fishes and macroinvertebrates. How the Marsh is managed, however, will determine what fishes are abundant.

APPENDIX 7.1 Fishes and macroinvertebrates of Suisun Marsh collected
in the University of California (Davis) fish study, 1979–2012.

Common name	Scientific name	Native/Alien
FISHES		
American shad	*Alosa sapidissima*	A
Bay pipefish	*Sygnathus leptorhynchus*	N
Bigscale logperch	*Percina macrolepida*	A
Black bullhead	*Ameiurus melas*	A
Black crappie	*Pomoxis nigromaculatus*	A
Bluegill	*Lepomis macrochirus*	A
Brown bullhead	*Ameiurus nebulosus*	A
California halibut	*Paralichthys californicus*	N
Chinook salmon	*Oncorhynchus tshawytscha*	N
Common carp	*Cyprinus carpio*	A
Delta smelt	*Hypomesus transpacificus*	N
Fathead minnow	*Pimephales promelas*	A
Golden shiner	*Notemigonus crysoleucas*	A
Goldfish	*Carassius auratus*	A
Green sturgeon	*Acipenser medirostris*	N
Green sunfish	*Lepomis cyanellus*	A
Hitch	*Lavinia exilicauda*	N
Largemouth bass	*Micropterus salmoides*	A
Longfin smelt	*Spirinchus thaleichthyes*	N
Longjaw mudsucker	*Gillichthys mirabilis*	N
Mississippi silverside	*Menidia audens*	A
Western mosquitofish	*Gambusia affinis*	A
Northern anchovy	*Engraulis mordax*	N
Pacific herring	*Clupea harengeus*	N
Pacific lamprey	*Lampetra tridentata*	N
Pacific sanddab	*Citharichthys sordidas*	N
Plainfin midshipman	*Porichthys notatus*	N
Prickly sculpin	*Cottus asper*	N
Rainwater killifish	*Lucania parva*	A
Redear sunfish	*Lepomis microlophus*	A
Sacramento blackfish	*Orthodon microlepidotus*	N
Sacramento pikeminnow	*Ptychocheilus grandis*	N
Sacramento splittail	*Pogonichthys macrolepidotus*	N
Sacramento sucker	*Catostomus occidentalis*	N
Shimofuri goby	*Tridentiger bifasciatus*	A
Shiner surfperch	*Cymatogaster aggregata*	N
Shokihaze goby	*Tridentiger barbatus*	A
Staghorn sculpin	*Leptocottus armatus*	N
Starry flounder	*Platichthys stellatus*	N
Steelhead	*Oncorhynchus mykiss*	N

(continued)

Common name	Scientific name	Native/ Alien
Striped bass	*Morone saxatilis*	A
Surf smelt	*Hypomesus pretiosus*	N
Threadfin shad	*Dorosoma petenense*	A
Threespine stickleback	*Gasterosteus aculeatus*	N
Tule perch	*Hysterocarpus traski*	N
Wakasagi	*Hypomesus nipponensis*	A
Warmouth	*Lepomis gulosus*	A
White catfish	*Ameiurus catus*	A
White crappie	*Pomoxis annularis*	A
White croaker	*Genyonemus lineatus*	N
White sturgeon	*Acipenser transmontanus*	N
Yellowfin goby	*Acanthogobius flavimanus*	A

MACROINVERTEBRATES

Black Sea jellyfish	*Maeotias marginata*	A
Siberian prawn	*Exopalaemon modestus*	A
Oriental Shrimp	*Palaemon macrodactylus*	A
California bay shrimp	*Crangon franciscorum*	N
Opposum shrimp	*Neomysis mercedis*	N
Overbite clam	*Corbula amurensis*	A
Asian clam	*Corbicula fluminea*	A
Harris mud crab	*Rhithropanopeus harrissii*	A

APPENDIX 7.2 Likely responses of selected fishes and macroinvertebrates in Suisun Marsh to changes in key variables resulting from sea-level rise and climate change.

Common name	Scientific name	Habitat	Present status	Higher salinity	Higher temperatures	More shallow open water	Less pelagic food	Likely population response
FISHES								
Delta smelt	Hypomesus transpacificus	P	Rare	--	--	+	--	Decrease
Longfin smelt	Spirinchus thaleichthys	P	Abundant	0	-	+	-	Decrease
Striped bass*	Morone saxatilis	P/L/B	Abundant	0	0	+	-	No change
Threadfin shad*	Dorosoma petenense	P	Abundant	--	+	+	-	Decrease
Chinook salmon	Oncorhynchus tshawytscha	P/L	Common	0	-	+	-	Decrease
Northern anchovy	Engraulis mordax	P	Rare	++	-	+	-	Increase
Mississippi silverside*	Menidia audens	P	Abundant	0	++	++	-	Increase
Black crappie*	Pomoxis nigromaculatus	L	Common	--	+	+	-	Decrease
Shiner perch	Cymatogaster aggregata	L	Rare	+	-	+	0	Increase
Threespine stickleback	Gasterosteus aculeatus	L	Abundant	0	+/-	+	0	No change
Sacramento splittail	Pogonichthys macrolepidotus	B	Abundant	0	0	+	0	Increase
White sturgeon	Acipenser transmontanus	B	Common	0	0	0	0	No change
White catfish*	Ameiurus catus	B	Abundant	--	+	0	0	Decrease
Staghorn sculpin	Leptocottus armatus	B	Common	+	0	+	0	Increase
Yellowfin goby*	Acanthogobius flavimanus	B	Abundant	0	+	0	0	Increase
Common carp*	Cyprinus carpio	B	Abundant	-	+	+	0	Decrease
Starry flounder	Platichthys stellatus	B	Common	++	0	+	0	Increase
Speckled sanddab	Citharichthys stigmaeus	B	Rare	+	0	0	0	Increase

(continued)

APPENDIX 7.2 *(continued)*

Common name	Scientific name	Habitat	Present status	Higher salinity	Higher temperatures	More shallow open water	Less pelagic food	Likely population response
MACROINVERTEBRATES								
Black Sea jellyfish*	*Maeotias marginata*	P	Abundant	+	+	0	-	Increase
Siberian prawn*	*Exopalaemon modestus*	P/B	Abundant	--	+	+	0	Decrease
Oriental Shrimp*	*Palaemon macrodactylus*	P/B	Common	+	?	+	?	Increase
California bay shrimp	*Crangon franciscorum*	P/B	Abundant	+	0	+	-	Increase
Opossum shrimp	*Neomysis mercedis*	P/B	Common	0	-	+	-	Decrease
Overbite clam*	*Corbula amurensis*	B	Abundant	+	+	0	-	Increase
Asian clam*	*Corbicula fluminea*	B	Abundant	-	+	0	-	Decrease

NOTE: Asterisk indicates alien species. Habitat abbreviations: P = pelagic, B = benthic, and L = littoral (shallow water edge). Response key: + = positive response, - = negative response, o = no response, and ++ and -- indicate especially strong responses.

REFERENCES

Bennett, W. A., and P. B. Moyle. 1996. Where have all the fishes gone: interactive factors producing fish declines in the Sacramento–San Joaquin estuary. Pages 519–542 *in* San Francisco Bay: The Ecosystem (J. T. Hollibaugh, ed.). San Francisco: AAAS Pacific Division.

Feyrer, F., B. Herbold, S. A. Matern, and P. B. Moyle. 2003. Dietary shifts in a stressed fish assemblage: consequences of a bivalve invasion in the San Francisco Estuary. Environmental Biology of Fishes 67: 277–288.

Feyrer, F., J. Hobbs, M. Baerwald, T. Sommer, Q. Yin, K. Clark, B. May, and W. Bennett. 2007a. Otolith microchemistry provides information complementary to microsatellite DNA for a migratory fish. Transactions, American Fisheries Society 136: 469–476.

Feyrer, F., M. L. Nobriga, and T. R. Sommer. 2007b. Multidecadal trends for three declining fish species: habitat patterns and mechanisms in the San Francisco Estuary, California, USA. Canadian Journal of Fisheries and Aquatic Sciences 64: 723–734.

Hatfield, S. 1985. Seasonal and interannual variation in distribution and population abundance of the shrimp *Crangon franciscorum* in San Francisco Bay. Hydrobiologia 129: 199–210.

Matern, S. A., P. B. Moyle, and L. C. Pierce. 2002. Native and alien fishes in a California estuarine marsh: twenty-one years of changing assemblages. Transactions of the American Fisheries Society 131: 797–816.

Meng, L., P. B. Moyle, and B. Herbold. 1994. Changes in abundance and distribution of native and alien fishes of Suisun Marsh. Transactions of the American Fisheries Society 123: 498–507.

Moyle, P. B., R. D. Baxter, T. Sommer, T. C. Foin, and S. A. Matern. 2004. Biology and population dynamics of Sacramento splittail (*Pogonichthys macrolepidotus*) in the San Francisco Estuary: a review. San Francisco Estuary and Watershed Science 2(2): Article 3.

Moyle, P. B., J. Hobbs, and T. O'Rear. 2012. Fishes. Pages 161–173 *in* Ecology, Conservation and Restoration of Tidal Marshes: The San Francisco Estuary (A. Palaima, ed.). Berkeley: University of California Press.

Oh, C., H. L. Suh, K. Park, and C. Ma. 2002. Growth and reproductive biology of the freshwater shrimp *Exopalaemon modestus* (Decapoda: Palaemonidae) in a lake of Korea. Journal of Crustacean Biology 22: 357–366.

O'Rear, T. A. 2012. Diet of an introduced estuarine population of white catfish in California. MS thesis, University of California, Davis.

O'Rear, T. A., and P. B. Moyle. 2010. Long term and recent trends of fishes and invertebrates in Suisun Marsh. Interagency Ecological Program Newsletter 23(2): 26–48.

O'Rear, T. A., and P. B. Moyle. 2011. Suisun Marsh fish study: trends in fish and invertebrate populations of Suisun Marsh January 2010—December 2010. California Department of Water Resources, Sacramento.

Schroeter, R. E. 2010. The temporal and spatial trends, biological constraints, and impacts of an invasive clams, *Corbula amurensis,* in Suisun Marsh, San Francisco Estuary. PhD dissertation, University of California, Davis.

Siegel, S., P. Bachand, D. Gillenwater, S. Chappel, B. Wickland, O. Rocha, M. Stephenson,

W. Heim, C. Enright, P. Moyle, P. Crain, B. Downing, and B. Bergamaschi. 2011. Final evaluation memorandum, strategies for reducing low dissolved oxygen and methylmercury events in northern Suisun Marsh. Prepared for the State Water Resources Control Board, Sacramento, California. SWRCB Project Number 06–283–552–0.

Siegfried, C. A. 1982. Trophic relations of *Crangon franciscorum* Stimson and *Palaemon macrodactylus* Rathbun: predation on the opossum shrimp, *Neomysis mercedis* Holmes. Hydrobiologia 89: 129–139.

Sommer, T., C. Armor, R. Baxter, R. Breuer, L. Brown, M. Chotkowski, S. Culberson, F. Feyrer, M. Gingras, B. Herbold, W. Kimmerer, A. Mueller-Solger, M. Nobriga, and K. Souza. 2007. The collapse of pelagic fishes in the upper San Francisco Estuary. Fisheries 32:270–277.

Wintzer, A. P., M. H. Meek, P. B. Moyle, and B. May. 2010. Ecological insights into the polyp stage of non-native hydrozoans in the San Francisco Estuary. Aquatic Ecology 45: 151–161.

Wintzer, A. P., M. H. Meek, and P. B. Moyle. 2011a. Life history and population dynamics of *Moerisia* sp., a non-native hydrozoan, in the upper San Francisco Estuary (U.S.A.). Estuarine, Coastal and Shelf Science 94: 48–55.

Wintzer, A. P., M. H. Meek, and P. B. Moyle. 2011b. Trophic ecology of two non-native hydrozoan medusae in the upper San Francisco Estuary. Marine and Freshwater Research 62: 952–961.

Xu, J., M. Zhang, and P. Xie. 2008. Stable isotope changes in freshwater shrimps (*Exopalaemon modestus* and *Macrobrachium nipponensis*): trophic pattern implications. Hydrobiologia 605: 45–54.

8

Suisun Marsh Today:
Agents of Change

Stuart W. Siegel

Suisun Marsh has changed, is changing, and will continue to change. The principal subject of this book—the future of Suisun Marsh—is about looking forward with the intention of directing that change toward a positive outcome in the future. The problem, of course, is that beauty is in the eye of the beholder, so one person's positive outcomes can be another person's bad dreams. Therefore, debate and dialogue among those who care about the Marsh are vital to looking forward with intention. Decisions about the future of the Marsh must be well informed if they are to have positive and predictable outcomes. Previous chapters have described historical changes to the Marsh, the status of its present flora and fauna, and physical processes that control ecological functions. Those chapters provide the starting point for planning the Marsh's future.

This chapter summarizes key agents of change: the diverse factors acting upon Suisun Marsh, their mechanisms, and further changes they are likely to bring about. We know more about some of these than we do others, resulting in varying degrees of speculation and uncertainty. To develop scenarios for alternative futures of the Marsh (see chapter 9), we need to understand the separate and combined effects of the following agents of change:

1. climate change
2. reduced sediment supply
3. invasive alien species
4. warmer temperatures
5. earthquakes and seismic risk
6. large-scale tidal marsh restoration

7. watershed land-use change
8. Delta water operations
9. salinity management
10. managed wetlands operations
11. management of endangered species
12. public policy and institutions

CLIMATE CHANGE

Climate change is anticipated to affect Suisun Marsh physically through three main mechanisms: sea-level rise, salinity intrusion, and storm frequency and intensity. Projections for all three mechanisms carry uncertainty, which increases as conditions further in the future are considered (Cayan et al. 2008, 2009). Projections are sensitive to complex atmospheric physical and chemical processes and to future human action or inaction to mitigate climate change.

Sea-level Rise

Sea-level rise is not new. Since 1850, mean sea level at the Golden Gate has risen approximately 0.25 m (0.8 ft) (see chapter 3). Projections for future climate-change-induced sea-level rise predict that the rate of rise will increase, and that the total rise will be 0.3–0.45 m (1–1.5 ft) by 2050 and 1.4 m (4.6 ft) by 2100 (Bay Conservation and Development Commission 2011; Bay Delta Conservation Plan [BDCP] 2013). Recent work comparing measured to projected sea-level rise suggests that the actual rise has been higher than predicted, which suggests a low bias in projections by the Intergovernmental Panel on Climate Change (IPCC) on which the local projections are based (Intergovernmental Panel on Climate Change 2007; Cayan et al. 2012; National Research Council 2012; Rahmstorf et al. 2012). Sea-level rise will continue beyond 2100 even if there is a drastic change in human interactions with the planet.

A key effect of sea-level rise is that the water surface will reach any given elevation with increased frequency, from daily low tides to extreme tide events, driving the 100-year tidal floodplain ever higher (figure 8.1). For example, what is today's 100-year flood stage could be the 1- or 5-year flood stage in the future.

For Suisun Marsh, sea-level rise can manifest its effects in many ways. First, it exacerbates problems created by subsidence. If land surface elevations are too low when areas are flooded, disturbance from wind-wave regimes will prevent sites from developing into emergent marsh. As sea level rises, additional flooded low-lying lands will be converted to open water rather than to emergent marsh (see map 9 in color insert and associated chart).

Higher projected salinities would add to this challenge, given plants' physiological limits to growth in saline intertidal waters (see chapters 3 and 4). Second, managed wetlands will face more frequent dike overtopping and be less able to drain by gravity. Third, tidal-marsh vegetation community succession will take place. Fourth, greater erosion along slough banks will affect dikes, increasing flooding of diked marshlands. Fifth, there will be a greater flood risk in surrounding communities and along Highways 12 and 680, and the Fairfield–Suisun City Wastewater Treatment Plant will have greater difficulty utilizing gravity drainage for its discharges.

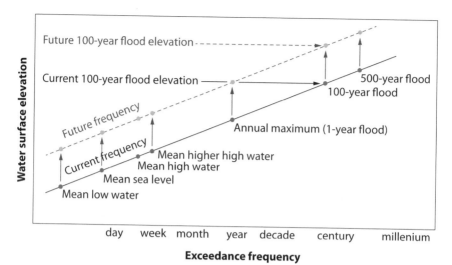

FIGURE 8.1. Diagram of tidal-water surface-elevation exceedance frequency today and with sea-level rise. *Exceedance frequency* is defined as how often a given event occurs (e.g., mean high water-surface elevation occurs once or twice a day, whereas the one-year flood event happens once a year) (source: Heberger et al. 2009).

Salinity Regime

Continuing sea-level rise will bring more saline ocean waters through the Golden Gate (chapter 3), resulting in higher Suisun Marsh salinities. Fall salinity increases may be most pronounced because fall is the driest time of year, and fall outflows of the Sacramento–San Joaquin Delta are unlikely to be increased by water operations to offset increased oceanic salinity. During winter and spring, shorter periods of low salinity are anticipated because precipitation will fall more as rain than as snow and arrive in fewer, more intense storms. With an overall increase in salinity, the Marsh will experience less spatial variability in salinity and could experience less intra- and inter-annual variability. If climate change creates more large flood events, however, high intra-annual variability could occur more often than today.

Increased salinity in Suisun Marsh will have several ramifications. Diked managed wetlands, which commonly are managed for low-salinity conditions, will have greater difficulty achieving those low salinities. Increased salinity will result in a plant community shift to more salt-tolerant species and reduced foraging habitat for most ducks and geese. A similar shift toward more salt-tolerant species would also occur in all other biotic communities, including aquatic communities (chapter 7). These shifts will ripple upward through the food web, altering the fish, wildlife, invertebrate, and plant species that the Marsh will be able to support. Key tidal salt-marsh species, such as salt marsh harvest mouse (*Reithrodontomys raviventris*) and California clapper rail (*Rallus longirostris obsoletus*), may benefit because conditions are likely to shift more toward the center of these species' ecological requirements.

Storm Frequency and Intensity

California is projected to retain its Mediterranean climate of cool and wet winters and warm and dry summers, including high variability in interannual precipitation amounts (Cloern et al. 2011). However, total precipitation (rain and snow combined) is generally projected to decline, or at least not increase, as is the total number of storms. While storms will likely be fewer, storm intensities are projected to increase, as are frequencies of extreme events (Cloern et al. 2011). Projections also suggest longer storm surges of salt water from the ocean (Cayan et al. 2008, 2009, 2012; Heberger et al. 2009). Even without climate change, storms that are orders of magnitude larger than ones experienced in recent decades, akin to the storm of 1861–62 when it rained for 43 days, will likely turn the Central Valley into a vast lake and make San Francisco Bay fresh; such storms have historically occurred once every 100–200 years (Dettinger and Ingram 2013).

For Suisun Marsh, the primary effects of these events would be increased flooding, by both salt water and fresh water, more levee damage, and greater intra-annual salinity variability. Longer and higher storm surges and more intense storm events will result in greater likelihood of overtopped dikes and of damage to those levees. To maintain the status quo, greater efforts to protect against, respond to, and repair storm damage will be needed, exacting an economic as well as an environmental cost. Eventually, the fight to maintain diked wetlands may be abandoned in many areas, resulting in tidal wetlands and open water (chapter 9).

REDUCED SEDIMENT SUPPLY

Sediment transport, supply, and deposition in Suisun Marsh have undergone major changes in the past 150 years (chapter 3). The first major change took place following the discovery of gold in the Sierra Nevada in 1848 with the commencement of hydraulic placer mining. This mineral extraction process released billions of cubic meters of sediment into the rivers and San Francisco Estuary, including Suisun Marsh—an estimated ninefold increase over premining levels (Gilbert 1917). Schoellhamer (2011) found that this pulse of sediment has now passed through its lowest estuarine position, San Pablo Bay, so projections today are of net sediment loss in the shallow bays of the Estuary. Although the primary pulse of mining sediment has moved through the Marsh system, remnant terrace deposits with some potential for erosion remain in many watersheds (Meade 1982; James 1991). While recent estimates of river sediment supply to the Delta are substantially higher than Gilbert's (1917) premining estimate, sediments have nevertheless continued to decrease since the mid-1950s, indicating ultimate exhaustion of remnant mining-derived deposits (Wright and Schoellhamer 2005).

The next major change in sediment supply resulted from the construction, beginning in the 1930s, of a vast network of dams and reservoirs for the Central Valley Project and State Water Project, alongside numerous smaller dams, throughout the Central Valley watershed. These dams collectively have trapped sediments in the upper watersheds. Although channels immediately downstream from the dams eroded to new equilibria (Porterfield et al. 1978), providing a short-term sediment source, the long-term effect has been to further decrease sediment supply to the Estuary (Williams and Wolman 1984). Dams also affected flow regimes, typically reducing high flows in most years and increasing low flows (Singer 2007), which together have the effect of reducing downstream sediment supply.

The reduction of hydraulic-mining sediment transport and construction of dams led to a 50% decrease in sediment supply from the Sacramento River between 1957 and 2004 (Wright and Schoellhamer 2004). Ganju and Schoellhamer (2010) suggested that sediment exchange between embayments of the San Francisco Estuary may become more significant sources of sediment as watershed sediment loads continue to decrease. The ability of those embayments to resuspend bottom sediments, however, is expected to decrease with increasing depth from sea-level rise (Schoellhamer 2011).

Schoellhamer (2011) provides evidence that the Estuary crossed a sediment supply threshold in 1999 when erodible sediment on its bottom became depleted. This sediment historically was resupplied by storm contributions from local watersheds and from Central Valley discharges flowing through the Delta, and then resuspended by tidal and wind-wave currents. The decline in external sediment inputs is expected to result in lower rates of sediment accretion in the Marsh. Increased water depths in restored wetlands due to sea-level rise increases the potential for wind-wave resuspension of deposited sediment. These two changes can preclude restoration sites from accreting sediment and produce low sedimentation rates.

This reduction of sediment supplies particularly affects estuarine marsh restoration sites. Most land suited for tidal-marsh restoration is former tidal marshland that subsided as a result of land-use practices. The subsidence reversal essential to reestablishing vegetated tidal marshlands and their ecological functions can come about through two mechanisms: accumulation of sediment deposited from the tidal water column, and accumulation of plant organic matter as both above- and belowground biomass. The reduced supply of suspended sediment for Suisun Marsh leaves organic-matter accumulation as a more critical mechanism for reversing subsidence. Plant biomass production is lower for saline-tolerant species, so shifts of the plant community to lower-productivity species in response to sea-level rise will lead to lower rates of organic-matter accumulation. The ultimate result is a much slower rate of subsidence reversal. Sea-level rise will exacerbate this problem. For more subsided restoration sites, where wind-wave resuspension outweighs sediment retention, these factors could act to keep them permanent open-water habitats. Active efforts to rebuild plant biomass prior to restoring tidal action may become an important design feature for restoration, especially for the more subsided sites.

Another effect of reduced sediment loads is increased water clarity and, thus, deeper light penetration into the water column. More light equates to a larger photic zone, which enables greater total phytoplankton productivity, potentially increasing support for aquatic food webs that lead to fish, assuming that the increased productivity is not all consumed by benthic bivalves. Greater light penetration can also allow increased growth of submerged aquatic vegetation, such as sago pondweed, benefiting waterfowl in shallow bays and sloughs (chapter 4).

INVASIVE ALIEN SPECIES

As shown in chapters 4, 6, and 7, Suisun Marsh now supports a large number of alien species; additional species are continually invading. Changes in hydrologic conditions will tend to favor already established alien species, especially those preferring estuarine or marine habitats. Predicting future invasions is possible to some extent (e.g., Lund et al. 2007), but there is a strong unpredictable component due to the rapid and continued movement of humans and commerce around the globe. The most definitive statement that can be made about future invasions is that they will occur unless stringent prevention measures are in place, and human intervention to avoid or control invasions may work in some cases but not in others.

Once a species invades the Marsh, its abundance and potential harmful effects depend on its ability to adapt to variable conditions. Wetland vegetation is sensitive to salinity during late winter–early spring germination periods, as evidenced by past periods of salinity changes (Clark and Patterson 1985; Clark 1986; Beare and Zedler 1987). Decreases in tidal-marsh salinity during seedling establishment can increase the spread of invasive plant species, such as the alien perennial pepperweed (*Lepidium latifolium*). Under normal conditions, salinity variation generally promotes dynamic plant communities by influencing interactions of dominant native perennial species and annuals or short-lived perennials that have adapted to varied levels of soil salinity (Callaway et al. 1990; Allison 1992). However, well-established invaders also can persist through less-than-favorable high-salinity conditions (Zedler 1983). Salinity regime also exerts an important influence on the abundance of alien invertebrates such as the overbite clam (*Corbula amurensis*) and four species of alien "jellyfish" (Hydrozoa) (Wintzer et al. 2011), as well as affecting the abundance of various fishes (chapter 7).

A persistent issue for restoration is the widespread extent of dominant alien plants (chapter 4) believed to interfere with desired ecological functions that are already established within a highly salinity-regulated marsh system. Such species are expected to be a continued problem in restoration areas, especially when established nearby. Successive years of high salinity stress are hypothesized to be an important factor controlling the spread and establishment of invasive plant species (Suisun Ecological Workgroup 2001).

In aquatic habitats of Suisun Marsh, biotic communities are mixtures of native and alien fishes and invertebrates, with little evidence of native species being driven to extinction by alien invaders. In fact, Matern and Brown (2005) could find little evidence that the invasion of shimofuri goby had harmful effects on any native species, despite its high abundance in the Marsh. In general, higher salinities also tend to favor native fishes (chapter 7). By contrast, the invasion of overbite clam caused major changes to food webs in the Estuary, including lower Suisun Marsh (Kimmerer et al. 1994; Feyrer et al. 2003), and its failure to invade the upper Marsh has increased the value of the Marsh as a refuge for native species (chapter 7).

Overall, Suisun Marsh biotic communities will continue to be mixtures of native and alien species. A basic management conundrum is how to favor desirable native species in the presence of aggressive invasive species, given that eradication of alien species is unlikely. Climate change is likely to create conditions more favorable to many alien species, as well as different assemblages of native species, requiring new and creative management strategies.

WARMER AIR AND SURFACE-WATER TEMPERATURES

Under virtually all realistic climate scenarios, the water in the Estuary and Marsh will have mean annual temperatures several degrees warmer than they are today. Cloern et al. (2011) projected the effects of climate change on the Delta using two very different models from the IPCC (2007) report. Scenario B1 is an optimistic scenario that assumes major reduction in greenhouse gases by 2050. Scenario A1 assumes continual increase in greenhouse gases, which is presumably more realistic but not necessarily the most extreme scenario possible. Under the B1 scenario, mean annual water temperatures would rise 1–2°C (to around 18°C) and the number of days when temperatures exceed 25°C would rise to 15–20 days per year. Under the A2 scenario, mean annual temperatures would rise 3–4°C (to around 20°C), with 80–100 days per year 25°C or above. Under A2, temperatures would keep rising sharply (Cloern et al. 2011). Although the moderating effect of rising sea level, wind, and ocean fog on temperatures in the Marsh is uncertain, temperatures are likely to rise to some extent, regardless. This means that many organisms that require cool water, such as delta smelt, will likely find the Marsh increasingly inhospitable from a temperature perspective.

EARTHQUAKES AND SEISMIC RISK

Suisun Marsh is underlain by a small number of seismic faults, located mostly around the margins of the Marsh (Graymer et al. 2002). Located along the western margin of the Marsh, the Green Valley fault, part of the Concord–Green Valley fault system, is the dominant mapped fault (Graymer et al. 2002) and appears to be the most active. The last large earthquake on this fault occurred 200–500 years ago, but a magnitude 3.2 earthquake occurred on October 8, 2012 (U.S. Geological Survey 2012), indicating that the fault is active. It is possible, even likely, that an earthquake large enough to shake down dikes in the Marsh will occur in the next 100 years, increasing the vulnerability of marshlands to sea-level rise.

LARGE-SCALE TIDAL MARSH RESTORATION

Large-scale tidal marsh restoration for Suisun Marsh in the near future seems increasingly likely. It is called for in the Suisun Marsh Plan (U.S. Bureau of Reclamation [USBR] et al. 2011), the BDCP, the Delta Plan, the Ecosystem Restoration Program (ERP) Stage 2 Conservation Strategy, the U.S. Fish and Wildlife Service (USFWS) San Francisco Estuary Tidal Marsh Recovery Plan, and the USFWS Delta Smelt Biological Opinion for water-project operations. The 2000 CALFED Record of Decision identified Suisun Marsh restoration to be in the range of 20 to 28 km^2 (5,000 to 7,000 acres). In the prior year, the Baylands Habitat Goals Report recommended 70 to 90 km^2 (17,000 to 22,000 acres). The Suisun Marsh Plan calls for the CALFED Record of Decision target, and BDCP and the Delta Plan may call for greater area.

Effects of Restoration

The primary effect of these proposed restoration projects, beyond benefits to native species, will be to alter tidal hydrodynamics within Suisun Marsh and beyond, by redirecting flow patterns and absorbing estuarine tidal energy (see chapter 3). Hydrodynamic modeling for the Suisun Marsh Plan has demonstrated that restoring tidal action to large areas of subsided lands absorbs significant tidal energy and reduces tidal range in nearby areas. The models suggest that mean low water may be up to 0.5 m higher than it is with the current channel configuration (USBR et al. 2011). Raising the elevation of low tides makes inter-tidal and subtidal lands lower in relation to the tides. The exact magnitude of this effect will depend on where tidal marsh restoration efforts are located. Areas that are less subsided will see less of an effect and will recover faster. As restoration sites fill in through mineral sedimentation and plant-matter accumulation, their tidal prisms will decrease and the effects of large-scale restoration on tidal ranges will diminish.

Large-scale tidal marsh restoration efforts can also have a variety of other effects, positive and negative. For example, they can promote invasive species, alter waterfowl distribution, and alter other ecosystem functions that support both resident and migratory species. Restoration can also promote the recovery of a large range of listed species. But changes caused by restoration projects in the Marsh will also be affected by large-scale changes to the Delta, such as flooding of subsided islands. These changes will absorb estuarine tidal energy and alter tidal ranges in the Marsh, the magnitude of effects depending on the location and timing of the changes. Together, the scale of Delta tidal island flooding and marsh restoration projects could be huge—in the many tens of thousands of acres. However, the role of the Marsh in this change will depend on the extent, timing, and geography of Delta restoration actions.

Regardless of Delta–Marsh interactions, wetland restoration projects will take considerable time to produce noticeable results. Tidal marsh restoration is an evolutionary process. Restoration sites take years to evolve from conditions on the day of a levee breach to a future "quasi-equilibrium" high marsh roughly akin to those on Brown's Island or Rush Ranch. In those intervening years, conditions typically change gradually, as do the associated ecological functions that a restoration site provides. Key step-changes can occur when process thresholds are crossed, such as when sediment accretion raises site elevations to heights where emergent vegetation can colonize. These complexities show why potential modifications to the Marsh should be placed in a comprehensive ecological framework that allows for a more nuanced approach to large-scale restoration.

Conceptual Models

A good way to understand factors that affect tidal marsh restoration is to develop conceptual models of potential interactions among factors. The initial conditions in the models presented here reflect baseline site elevation, substrate characteristics, and the composition of emergent vegetation (Siegel et al. 2010). Once diked lands are opened to tidal action, the many physical and biological processes that control site evolution take over. Ecological functions are tied very strongly to the progress of a restoration site along this evolutionary trajectory (Siegel et al. 2010). Figure 8.2 shows the relationship between Suisun Marsh elevation and inundation regime, which exerts a major influence over all aspects of site ecology.

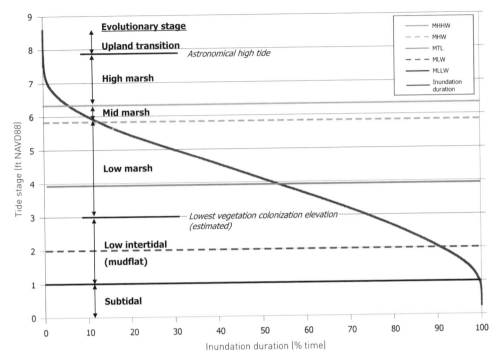

FIGURE 8.2. Inundation regime, marsh elevation, and habitat trajectories. Inundation duration (cumulative percentage of total time of submergence) is shown as a function of tide stage (curved line). When applied to tidal restoration, the elevation of the initial restoration site dictates the extent of tidal submergence, which controls habitat types and ecological functions. The lowest tidal elevation of emergent vegetation is driven by salinity and hydroperiod and is lower for freshwater species. In the Marsh, the lowest elevation of vegetated tidal marsh may extend a little below mean tide level. As a restoration site regains elevation through sediment and biomass accumulation, its vegetation communities, amounts of open water vs. vegetation, and ecological functions will shift upward through the habitat stages (subtidal to low intertidal to low marsh, etc.) (source: Siegel et al. 2010).

The rate of evolution is controlled predominantly by the relative influences and magnitudes of the four major drivers of restoration outcome: initial elevation (figure 8.2), hydrodynamic energy (chapter 3), sediment supply (chapter 3), and salinity and its associated control of vegetation community composition (chapter 4). Table 8.1 summarizes the directionality that each of these four drivers exerts upon the evolution of tidal marsh restoration.

Figure 8.3 illustrates processes that affect the rate at which tidal marsh restoration sites evolve toward high-elevation marsh. The interactions of these processes,

TABLE 8.1 Drivers of the rate (faster or slower) at which tidal marsh restoration evolves.

Driver	High magnitude	Low magnitude
Sediment supply	More sediment = faster	Less sediment = slower
Hydrodynamic energy	More energy = slower	Less energy = faster
Initial elevation	Higher elevation = faster	Lower elevation = slower
Salinity and vegetation	Higher salinity = slower	Lower salinity = faster

and the resulting rates of marsh accretion, control the range of ecological functions provided by restoration efforts. Figure 8.3 illustrates the general relationships among the many physical and biological processes and characteristics and
the ecological functions provided by tidal marshlands. Together these conceptual
models inform how restoration efforts will evolve as geomorphic elements of
Suisun Marsh and how they will provide a range of ecological functions.

WATERSHED LAND-USE CHANGE

Growth of Fairfield and Suisun City continues, as does development on unincor-
porated county lands to the west and north of Suisun Marsh, converting open
space, abandoned industrial sites, and agricultural lands to urban and industrial uses.

The primary effects of local land-use change are increases in impervious
surfaces leading to greater storm-water runoff that drains untreated to Suisun
Marsh, an increase in treated wastewater discharge, and a change in types and
amounts of nonpoint-source contaminants. In addition, more residences located
near the Marsh increase the demand for mosquito control, which can be accom-
plished through water management in the diked managed wetlands, hydrologic
modifications in the tidal marshes, and treatment with approved chemicals—all
of which have ecological implications.

Farther upstream in the watershed are agricultural and open lands, including annual and perennial crops and cattle grazing. The main changes in these
upstream land uses will likely be in the type, quantity, and timing of fertil-
izer, herbicide, fungicide, and insecticide applications. These changes will bring
increased volumes of treated wastewater discharge from the Fairfield–Suisun
City Wastewater Treatment Plant, adjacent to the northwest Marsh. That facility
operates part of the year without discharge to the Marsh, via field irrigation to
support cattle feed. Additionally, the facility is permitted to discharge up to 16
million gallons per day and has infrastructure to discharge tertiary-treated efflu-
ent into three diked managed wetlands and into Boynton and Peytonia sloughs.
The cumulative consequences of these land-use changes include increased loads
of a wide range of nonpoint-source pollutants and wastewater constituents, small

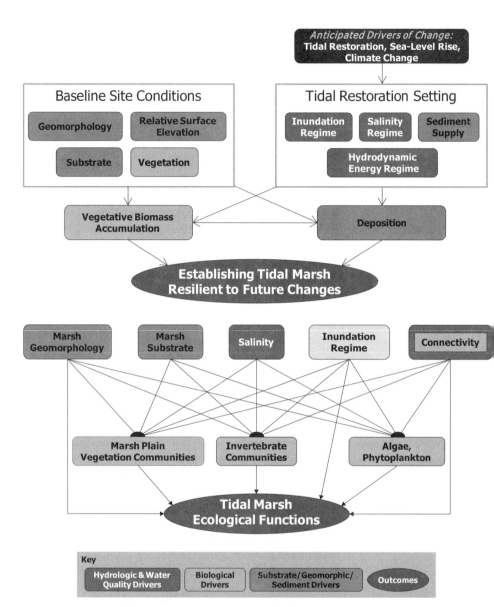

FIGURE 8.3. Conceptual models. (Top) Factors influencing tidal marsh restoration. Where a site "starts" (baseline conditions) and its position in relation to physical and biological processes (restoration setting) largely control how restoration progresses toward higher-elevation tidal marsh and may maintain functions as future changes arrive. (Bottom) Factors influencing tidal-marsh ecological functions. Physical processes and connectivity (upper boxes) control the biology (middle boxes) that yield target tidal-marsh ecological functions (bottom box). (Siegel et al. 2010).

increases in dry-season runoff from landscape irrigation, and loss of adjacent open habitats used by a variety of wildlife species.

DELTA WATER OPERATIONS

Water diversions from the Delta and its inflowing rivers cause changes in the complex flow dynamics in the Delta and Suisun Marsh that affect migration and movement of fish and other aquatic organisms, limit access to suitable habitats, and alter water quality. Together, the state and federal export facilities in the south Delta pump approximately 7.4 billion m^3 (6 million acre-feet) of water annually, and at times effectively reverse flows in the Old and Middle rivers, two major outflow channels in the south Delta. Other water diversions and consumptive water use within the Delta (over 2,000 diversions) use an additional 2.1 billion m^3 (1.7 million acre-feet) (Bay Institute 1998; Herren and Kawasaki 2001; Healey et al. 2008). Variability in flow regime and seasonal flooding were historically important drivers of ecosystem structure and processes in the Delta and the Marsh. Native plant and animal species evolved under flow regimes of high inter- and intra-annual variability that differed strongly from the current managed regime. Water storage and flow control have dampened such variability across seasons and years, which has greatly changed estuarine hydrodynamics, circulation patterns, and nutrient exchanges and has negatively affected resident species adapted to this natural variability (Moyle et al. 2010).

As a result of the severe ecological impacts that have resulted from large-scale water diversions and replumbing of the Delta, three key efforts aim to remedy these conflicts: the BDCP, the Delta Plan, and revisions to flow criteria that are part of the State Water Resources Control Board's water-rights agreements. These efforts have numerous political and scientific complexities and are unlikely to result in dramatic changes to existing Delta outflow regimes in the short term. However, as these regulatory efforts mature into projects that are implemented on the ground, the status quo could very well change. The stated intent of these efforts, as embodied in the Delta Reform Act of 2009, is to both provide a reliable water supply and restore the ecological integrity of the Delta, in which Suisun Marsh is often included.

The mostly likely changes to water operations will be in the magnitude, duration, and timing of Delta outflow, a major driver of physical processes and ecological functions in the Marsh (see chapter 3). These changes can affect vegetation community composition in the tidal marsh, occurrence and control of invasive species, aquatic food-web productivity, operations of diked managed wetlands, water-quality suitability of the Marsh for fish and aquatic organisms, sediment supply, and, consequently, the ability of ecological restoration activities to yield benefits to listed species and natural communities.

SALINITY MANAGEMENT

Salinity management within Suisun Marsh (described in chapter 3) is predicated on the maintenance of primarily freshwater conditions during winter and spring to facilitate managed duck hunting operations. Management is achieved by managing Delta outflows, operating salinity management infrastructure (especially large tidal gates located in Montezuma Slough), and controlling water operations in individual managed wetlands. Salinity will change in the Marsh as (1) Delta water operations introduce a new flow regime; (2) sea-level rise brings oceanic waters farther up the Estuary; (3) tidal marsh restoration alters hydrodynamics; and (4) flooded islands, tidal marsh restoration, and other changes in the Delta alter hydrodynamics. The future component most difficult to predict is human response to these changes. Will Suisun Marsh's salinity management infrastructure continue to be operated to maintain current conditions? Will new regimes be established as these changes come into play? Changes in magnitude and timing of salinity regimes, and associated human responses, will likely determine the magnitude of subsequent ecological change observed in existing and restored tidal marshes, in tidal aquatic ecosystems, and in associated species and natural communities that use these ecosystems (see chapter 3).

MANAGED WETLANDS OPERATIONS

The Suisun Marsh Plan (USBR et al. 2011) prescribes a range of actions for managing Suisun Marsh over the next 30 years. A majority of these actions fall into two categories: managed wetland operations and tidal marsh restoration. The plan calls for restoration of 20–28 km^2 (5,000–7,000 acres) of diked lands to tidal marsh; enhancement of 160–200 km^2 (40,000–50,000 acres) of diked, managed wetlands through improving interior water circulation and exterior water exchange; and allowing a return to tidal slough dredging for sediment to maintain diked marsh levees. These actions, upon receipt of all necessary regulatory authorizations, are designed to allow for a continuation of diked wetlands management over a 30-year period, consistent with operations and waterfowl habitat objectives in place for the past several decades. Duck club managers assume that these actions will take place, allowing continuation of the status quo for duck club marshland management (chapter 5).

Diked wetlands in Suisun Marsh are managed mostly to facilitate production of resident waterfowl and to attract and support migratory waterfowl for hunting; thus, they are managed as seasonally inundated, nontidal wetlands. Nevertheless, diked wetlands also provide a range of habitats that benefit many desirable native species, contributing greatly to diversity and ecological function. For example, several areas, totaling 10 km^2 (2,500 acres), are managed explicitly to provide habitat for the federally protected salt marsh harvest mouse (chapter 6).

Unfortunately, management of current diked wetlands also has adverse effects. Drying of seasonal wetlands causes peat soils to oxidize, resulting in land subsidence. Soil oxidation also produces carbon dioxide, contributing to greenhouse gas emissions. Ongoing subsidence increases the difficulty of managed wetlands operations because it makes gravity drainage less effective. It also increases the difficulty of restoration by increasing the need for subsidence reversal. Prior to the 1990s, spoils from dredging of tidal sloughs were used for levee maintenance. When regulations to protect endangered fishes prohibited this dredging, many managed wetlands borrowed soil internally in large quantities in order to maintain their levees. This practice has greatly lowered internal elevations. Diked wetlands also modify wetland geomorphology in ways that are designed to benefit their management but may make subsequent tidal marsh restoration more difficult. Seasonal wetland management in portions of Suisun Marsh also causes depletion of dissolved oxygen in sloughs that receive drainage water from the diked marshes, through the anoxic decomposition of plant matter and soil carbon. These same conditions also produce methylmercury (Siegel et al. 2011). Low-dissolved-oxygen events most commonly occur during "fall flood-up," when the wetlands shift from dry to saturated conditions, and to some degree in late winter during flood-and-drain cycles to leach salt from soils. In the smaller sloughs, these discharges limit habitat for many aquatic species and, in severe instances, result in fish kills. Methylmercury, a neurotoxin, can accumulate in the aquatic food web (Alpers et al. 2008) and result in human exposure through fish consumption. Suisun Marsh is one of many locations in California with posted public health warnings about methylmercury and fish consumption. The ecotoxicological effects of methylmercury exposure are less clearly understood (Alpers et al. 2008).

Ongoing managed wetlands operations thus present several consequences for the future of Suisun Marsh. Benefits to waterfowl are high. However, management for waterfowl needs to be more compatible with reducing or stopping soil subsidence, with maintaining good water quality in tidal sloughs that support diverse fish populations, and with decreasing creation of methylmercury. The most challenging effects of ongoing managed wetlands operation are subsidence and soil salt leaching. Subsidence will, over time and in combination with sea-level rise, create less effective gravity drainage and increase the need for pumps. Soil salt leaching, if it continues its reliance on low-salinity applied water, will become less practicable as surface water salinities increase with sea-level rise and, perhaps, with changes to Delta water operations. Shifts in the vegetation community to more salt-tolerant plant species will also result. Thus, continuation of current management practices will ultimately lead to less desirable conditions for waterfowl in diked wetlands.

MANAGEMENT OF ENDANGERED SPECIES

Suisun Marsh has a fairly long history of accommodating species listed under state and federal endangered species acts (ESAs) (chapters 6 and 7). Key ESA drivers of landscape-scale management have included salt marsh harvest mouse, California clapper rail, delta smelt, and salmonids and other fishes; several other species, including some plants, are being considered.

Approximately 10 km^2 (2,500 acres) of diked wetlands in the Marsh are geared toward recovery of the salt marsh harvest mouse. Were these lands to be considered for tidal marsh restoration, their mouse habitat function would have to be reestablished elsewhere in the Marsh. Importantly, the Marsh is on the eastern fringe of this species' range. Thus, because habitat quality would be marginal under natural conditions, most conservation is done on diked rather than tidal marsh, where more suitable conditions can be carefully maintained.

Protections for resident and migratory fishes have been major drivers of fish-screen installation, diversion restrictions on diked wetlands, and the cessation of slough dredging for diked wetland levee maintenance. The Suisun Marsh Plan, when fully implemented, will allow slough dredging for levee maintenance, thereby reducing or eliminating degradation of wetland interiors caused by large-scale borrowing of soils from island interiors. Recovery actions for listed fish species can support overall ecosystem restoration and conservation, as long as efforts are not focused solely on the listed species, but instead on ecosystem functions that support a broad spectrum of wildlife. Suisun Marsh is well suited to this purpose because it contains so many manageable features compared with other locations in the Estuary and, especially, the Delta.

PUBLIC POLICY AND INSTITUTIONS

This final section examines the current and proposed public policy arrangements that do or will govern changes in Suisun Marsh management, and how they support or hinder effective change.

Water-Project Biological Opinions

In 2008 and 2009, the USFWS and the National Marine Fisheries Service issued Biological Opinions under the federal ESA for joint operation of the State Water Project and federal Central Valley Project. In 2009, the California Department of Fish and Game (now Department of Fish and Wildlife) issued an Incidental Take Permit under the California ESA for the water projects. Though these actions require a total of 32.3 km² (8,000 acres) of tidal marsh restoration in the Delta and Suisun Marsh, only the California Department of Fish and Wildlife's requirement identifies 3.2 km² (800 acres) specific to the low-salinity zone of the Estuary. These requirements, all directed to listed fish species (delta smelt, salmon, green sturgeon, and longfin smelt), are intended to "restore tidal and associated subtidal habitats." An obvious target for these restoration efforts is Suisun Marsh, with the likelihood of considerable financial resources being brought to Marsh projects as a consequence. In the near term, fulfillment of these ESA requirements will provide the dominant funding mechanism and implementing entities for tidal marsh restoration in Suisun Marsh.

Bay Delta Conservation Plan

Currently under development, the BDCP would provide the environmental conservation measures associated with issuance of 50-year permits for Delta water exports. Suisun Marsh is identified as one of several geographic "restoration opportunity areas" (BDCP 2013). Similar to the Biological Opinions, the BDCP allows for tidal emergent marsh and tidal aquatic restoration. Although BDCP has a strong fish focus, it also provides some attention to the broader ecosystem. It would support restoration in the Marsh but does not bring financial resources to the table at this time, because its current financing plan relies on future voter-approved general obligation bonds for which the likelihood of passage is unpromising. Were those bonds passed, the entities that would apply these restoration funds are not yet clear, but they may include the Delta Conservancy, Department of Water Resources, Department of Fish and Wildlife, and/or some other as-yet undetermined entity.

Suisun Marsh Plan and Suisun Marsh Preservation Agreement

The Suisun Marsh Habitat Management, Preservation, and Restoration Plan (USBR et al. 2011) provides the current policy framework and environmental, ESA and regulatory compliance for long-term managed wetlands operations and implementation of the regulatory obligations of the state and federal water projects in Suisun Marsh that are incorporated into the Suisun Marsh Preservation Agreement. It also provides environmental assessment, but not regulatory compliance, for tidal marsh restoration of up to 28 km² (7,000 acres) of tidal marsh. The Suisun Marsh Plan supports actions that protect listed species and natural communities and that facilitate tidal marsh restoration for a broad range of benefits. However, it also perpetuates and funds management actions, some of which may be detrimental to other ecological functions of the Marsh (e.g., seasonal wetting and drying of marshlands, resulting in subsidence). Though the Suisun Marsh Plan includes some requirements for tidal marsh restoration, funding to fulfill those requirements is not included.

Suisun Marsh Protection Act

Passed in 1977, the Suisun Marsh Protection Act established the basis for Suisun-area conservation through land-use regulation implemented by the San Francisco Bay Conservation and Development Commission in the wetlands and by Solano County in the adjacent uplands. The act established managed wetlands as the primary land use in the Suisun Marsh interior, allowed for exploration and development of natural gas reserves, allowed for the Potrero Hills Landfill, and set aside the Collinsville area as an industrial reserve area. The latter has recently been reduced in size from 10.5 km^2 (2,600 acres) to 0.8 km^2 (200 acres). The Suisun Marsh Protection Act and its implementing regulatory programs are generally supportive of ecological restoration and conservation in Suisun, although some of the implementation policies are outdated and may not adequately accommodate sea-level rise and conservation of the wetland–upland transition zone important to so many species (chapter 4).

Clean Water Act and California Water Rights Decisions

Suisun Marsh water quality is regulated under a number of programs and plans. Salinity has long been regulated, beginning with Water Rights Decision D-1485 in 1978, Decision D-1641 in 1999, and three Water Quality Control Plans of 1978, 1995, and 2006. The San Francisco Bay Regional Water Quality Control Board (RWQCB) regulates beneficial uses of Suisun waterways through its Basin Plan. The RWQCB also maintains the CWA Section 303(d) Impaired Water Bodies list, which, for the Marsh, includes various pesticides, various organic compounds, mercury, selenium, low dissolved oxygen, nutrients, and marsh salinity. As part of the impaired water bodies program, the San Francisco Bay RWQCB (2006) has developed, and the U.S. Environmental Protection Agency has adopted, a mercury Total Maximum Daily Load (TMDL) for San Francisco Bay that covers the Marsh. Collectively, these water-quality criteria and plans establish a complex regulatory overlay that generally supports ecological restoration and conservation in Suisun, though they may pose challenges where restoration and conservation efforts conflict with salinity criteria. The San Francisco Bay RWQCB is currently in the process of developing low-dissolved-oxygen and methylmercury TMDL standards for Suisun Marsh.[1]

ERP Stage 2 Conservation Strategy, Delta Plan, and USFWS Tidal Marsh Recovery Plan

The ERP Stage 2 Conservation Strategy for the Delta (DFG 2011) was written specifically in anticipation of impending changes to how water is conveyed to the

---

1. See http://www.waterboards.ca.gov/.

state and federal water export facilities in the southern Delta. This Conservation Strategy will guide the Ecosystem Restoration Program Stage 2 implementation in the Delta and Suisun planning area and is incorporated into the Delta Plan adopted by the Delta Stewardship Council in 2013. The Delta Stewardship Council also has a quasi-regulatory role wherein restoration-project proponents must file "consistency determinations" demonstrating how the project is consistent with the Delta Plan through application of "best available science." The Delta Science Program is anticipated to have a role in supporting development and application of best available science. Efforts to develop a Delta Science Plan are underway as of the publication of this book. The Draft Tidal Marsh Recovery Plan (USFWS 2010) is a formal ESA-guided plan to chart recovery of tidal marsh–dependent species from Suisun Marsh through the San Francisco Estuary and along the Central California Coast. It contains specific restoration recommendations and targets for Suisun Marsh. Taken collectively, these three plans and programs support ecosystem conservation in Suisun Marsh and provide strong scientific foundations for restoration efforts.

Oversight Gap

There is an absence of clear statements on the leadership and authority of the many agencies that have responsibility in Suisun Marsh management. Under the Suisun Marsh Plan, tidal marsh restoration is "to happen," and whichever entities choose to pursue it have a prescribed set of procedures to follow in planning in order to utilize the environmental coverage the plan provides. The Suisun Marsh Plan establishes an Adaptive Management Advisory Team (AMAT), but it does not prescribe any linkage to science or adaptive management of the Delta Plan and the Delta Science Program that are the repository and guiding body for the vast stores of scientific knowledge on Suisun Marsh. The Delta Plan, adopted in 2013, mandates science engagement with the Delta Science Program through its quasi-regulatory "Covered Actions" consistency determinations; restoration in Suisun Marsh, for example, is a Covered Action. The Delta Science Plan, which is anticipated to be established in 2014, will provide guidance on the means by which the best available science should be incorporated into restoration efforts. The AMAT has not, to date, formed or prescribed procedures for establishing projects and designs within a strong science-based framework, nor announced how it will meet Delta Plan Covered Action requirements. The Suisun Marsh Plan's directed adaptive management program is geared mainly toward objectives that are a subset of larger policy directives described above. The plan does identify reducing uncertainties as an important priority. Adaptive management monitoring is defined for impacts of the managed-wetland levee-maintenance dredging program. Monitoring restoration benefits to listed species is deemed "potential," and no monitoring is geared toward ecosystem-level benefits. The plan does not establish a science oversight entity on par with the Delta Independent Science Board. Instead, it recommends that restoration projects seek "input of other science based work groups . . . as applicable." Such issues need to be resolved before effective adaptive management can happen.

SUMMARY AND CONCLUSIONS

Many and varied forces will continue to be acting upon the Suisun Marsh of the future, which should place land managers and resource planners in an adaptive management role. Our best opportunity is to plot futures of Suisun Marsh that take advantage of these forces where possible, and design a system that is as resilient to change as possible. For example, the Marsh may become well suited to interim wetland uses such as carbon sequestration, helping in a small way to mitigate climate change while preparing the Marsh for more effective restoration to tidal marsh through early subsidence reversal. Other forces are within our grasp to alter if we choose (chapter 9). In practice, it will be very challenging, if not impossible, to retain some of today's ecosystem functions and physiographic features, because we will be dealing with novel ecosystems (chapter 9).

Chapter 3 notes that the most critical challenges for tidal marsh restoration in Suisun Marsh are sea-level rise, sediment supply, land subsidence, and, to a lesser extent, tidal energy. Though sea-level rise and sediment supply are outside our control, two forces are fully within our control: land subsidence and tidal energy distribution. Land subsidence, an effect of diked wetland management, can be addressed through alteration of diked wetland management. Understanding tidal energy distribution, which would be altered by planned restoration efforts, requires complex two- and three-dimensional hydrodynamic models to examine a wide range of possible Marsh-wide restoration scenarios. These tools exist, and the challenge is to describe a number of different alternatives, each reflecting a mix of large-scale tidal marsh restoration in the Delta, the Marsh, and northern San Francisco Bay, in combination with options for Delta water conveyance and floodplain enhancement in Yolo Bypass and the San Joaquin River.

To move forward, we need several ingredients. We need to apply our best technical savvy, including the range of information presented throughout this book, to execute tidal marsh restoration, improve diked wetlands management, and improve water quality. We need the many public institutions involved in Suisun Marsh management to function effectively with leadership, authority, and responsibility. We need adequate funding with a reasonable degree of certainty. Perhaps most importantly, we need private landowners to become partners in charting the long-term future of Suisun Marsh. Many changes will be coming in the years and decades ahead. As illustrated in the following chapter, what the future Suisun Marsh looks like will depend on actions taken today.

REFERENCES

Allison, S. K. 1992. The influence of rainfall variability on the species composition of a northern California salt marsh plant assemblage. Vegetation 101:145–60.

Alpers, C. N., C. Eagles-Smith, C. Foe, S. Klasing, M. C. Marvin-DiPasquale, D. G. Slotton, and L. Windham-Myers. 2008. Mercury ecosystem conceptual model. Delta Regional Ecosystem Restoration Implementation Plan. http://www.science.calwater.ca.gov/drerip/drerip_index.html.

Bay Conservation and Development Commission. 2011. Living with a Rising Bay: Vulnerability and Adaptation in San Francisco Bay and on its Shoreline. Approved on October 6. San Francisco Bay Conservation and Development Commission, San Francisco.

Bay Delta Conservation Plan. 2013. Bay Delta Conservation Plan. Revised administrative draft. March 2013. ICF 00343.12. http://baydeltaconservationplan.com/Library/DocumentsLandingPage/BDCPDocuments.aspx.

Bay Institute. 1998. From the Sierra to the Sea: The Ecological History of the San Francisco Bay–Delta Watershed. Novato, California: Bay Institute.

Beare, P. A., and J. B. Zedler. 1987. Cattail invasion and persistence in a coastal salt marsh: the role of salinity reduction. Estuaries and Coasts 10: 165–170.

California Department of Fish and Game. 2011. Draft Ecosystem Restoration Program Conservation Strategy for Restoration of the Sacramento–San Joaquin Delta Ecological Management Zone and the Sacramento and San Joaquin Valley Regions. California Department of Fish and Game, Sacramento.

California Department of Water Resources. 1995. Sacramento–San Joaquin Delta Atlas (July). California Department of Water Resources, Sacramento.

Callaway, R. M., S. Jones, W. R. Ferren, Jr., and A. Parikh. 1990. Ecology of a Mediterranean-climate estuarine wetland at Carpenteria, California: plant distributions and soil salinity in the upper marsh. Canadian Journal of Botany 68: 1139–1146.

Cayan, D. R., P. D. Bromirski, K. Hayhoe, M. Tyree, M. D. Dettinger, and R. E. Flick. 2008. Climate change projections of sea level extremes along the California Coast. Climatic Change 87 (Supplement 1): S57–S73.

Cayan, D., M. Tyree, M. Dettinger, H. Hidalgo, T. Das, E. Maurer, P. Bromirski, N. Graham, and R. Flick. 2009. Climate change scenarios and sea level rise estimates for the California 2008 Climate Change Scenarios Assessment. Draft Paper, California Energy Commission CEC-500-2009-014-D.

Cayan, D. M. Tyree, D. Pierce, and T. Das. 2012. Climate Change and Sea Level Rise Scenarios for California Vulnerability and Adaptation Assessment. Publication No. CEC-500-2012-008. California Energy Commission, Sacramento.

Clark, J. S. 1986. Late-holocene vegetation and coastal processes at a Long Island tidal marsh. Journal of Ecology 74: 561–578.

Clark, J. S., and W. A. Patterson III. 1985. The development of a tidal marsh: upland and oceanic influences. Ecological Monographs 55: 189–217.

Cloern, J. E., N. Knowles, L. R. Brown, D. Cayan, M. D. Dittinger, T. L. Morgan, D. H. Schoellhamer, M. T. Stacey, M. van der Wegen, R. W. Wagner, and A. D. Jassby. 2011. Projected evolution of California's San Francisco bay-delta-river system in a century of climate change. PLoS ONE 6(9): e24465. doi:10.1371/journal.pone.0024465.

Cohen, A.C., and J.T. Carlton. 1995. Nonindigenous aquatic species in a United States estuary: a case study of the biological invasions of the San Francisco Bay and Delta. U.S. Fish and Wildlife Service and National Sea Grant Report No. PB96-166525.

Delta Stewardship Council. 2012. Preliminary Staff Review Draft. Final Staff Draft Delta Plan. Delta Stewardship Council, Sacramento.

Dettinger M.D., and D.L. Ingram. 2013. The coming megafloods. Scientific American 398(1): 64-71.

Feyrer, F., B. Herbold, S.A. Matern, and P.B. Moyle. 2003. Dietary shifts in a stressed fish assemblage: consequences of a bivalve invasion into the San Francisco Estuary. Environmental Biology of Fishes 67: 277-288.

Ganju, N.K., and D. Schoellhamer. 2010. Decadal-timescale estuarine geomorphic change under future scenarios of climate and sediment supply. Estuaries and Coasts 33: 1559-2723.

Gilbert, G.K. 1917. Hydraulic mining debris in the Sierra Nevada. U.S. Geological Survey Professional Paper 105.

Graymer, R.W., D.L. Jones, and E.E. Brabb. 2002. Geologic map database of northeastern San Francisco Bay region, California. U.S. Geological Survey Miscellaneous Field Studies Map MF-2403.

Healey, M.C., M.D. Dettinger, and R.B. Norgaard, eds. 2008. The State of the Bay–Delta Science. CALFED Science Program, Sacramento.

Heberger, M., H. Cooley, P. Herrara, P.H. Gleick, and E. Moore. 2009. The impacts of sea level rise on the California Coast. California Climate Change Center, CEC-500-2009-024-F. Pacific Institute, Oakland, California. http://www.pacinst.org/reports/sea_level_rise/report.pdf.

Herren, J.R., and S.S. Kawasaki. 2001. Inventory of water diversions in four geographic areas in California's Central Valley. Pages 343-355 in Contributions to the Biology of Central Valley Salmonids, vol. 2 (R.L. Brown, ed.). California Department of Fish and Game Fish Bulletin No. 179.

Intergovernmental Panel on Climate Change. 2007. Climate Change 2007—Impacts, Adaptation and Vulnerability. Contribution of Working Group II to the Fourth Assessment Report of the IPCC. Cambridge: Cambridge University Press.

James, L.A. 1991. Incision and morphologic evolution of an alluvial channel recovering from hydraulic mining sediment. Geological Society of America Bulletin 103: 723-736.

Kimmerer, W.J., E. Gartside, and J.J. Orsi. 1994. Predation by an introduced clam as the likely cause of substantial declines in zooplankton of San Francisco Bay. Marine Ecology and Progress Series 113: 81-93.

Lund, J., E. Hanak., W. Fleenor, W., R. Howitt, J. Mount, and P. Moyle. 2007. Envisioning Futures for the Sacramento–San Joaquin Delta. Public Policy Institute of California, San Francisco.

Matern, S.A., and L.A. Brown. 2005. Invaders eating invaders: exploitation of novel alien prey by the alien shimofuri goby in the San Francisco Estuary, California. Biological Invasions 7: 497-507.

Meade, R.H. 1982. Sources, sinks, and storage of river sediment in the Atlantic drainage of the United States. Journal of Geology 90: 235-252.

Moyle, P. B., W. A. Bennett, W. E. Fleenor, and J. R. Lund. 2010. Habitat variability and complexity in the upper San Francisco Estuary. San Francisco Estuary and Watershed Science 8(3): 1–24. http://escholarship.org/uc/item/okf0d32x.

National Research Council. 2012. Sea-Level Rise for the Coasts of California, Oregon, and Washington: Past, Present, and Future. Washington, D.C.: National Academies Press.

Porterfield, G., R. D. Busch, and A. O. Waananen. 1978. Sediment transport in the Feather River, Lake Oroville to Yuba City, California. U.S. Geological Survey Water-Resources Investigations 78–20: 73.

Rahmstorf, S., G. Foster, and A. Cazenave. 2012. Comparing climate projections to observations up to 2011. Environmental Research Letters 7: 044035. doi:10.1088/1748-9326/7/4/044035.

Sacramento–San Joaquin Delta Conservancy. 2012. 2012 Strategic Plan. Delta Conservancy, West Sacramento, California.

San Francisco Bay Regional Water Quality Control Board. 2006. Mercury in San Francisco Bay: proposed basin plan amendment and staff report for revised total maximum daily load (TMDL) and proposed mercury water quality objectives. San Francisco Bay Regional Water Quality Control Board, Oakland, California.

Schoellhamer, D. H. 2011. Sudden clearing of estuarine waters upon crossing the threshold from transport to supply regulation of sediment transport as an erodible sediment pool is depleted: San Francisco Bay, 1999. Estuaries and Coasts 34: 885–899.

Siegel, S., P. Bachand, D. Gillenwater, S. Chappell, B. Wickland, O. Rocha, M. Stephenson, W. Heim, C. Enright, P. Moyle, and others. 2011. Final evaluation memorandum, strategies for resolving low dissolved oxygen and methylmercury events in northern Suisun Marsh. SWRCB Project No. 06-283-552-0. Prepared for the State Water Resources Control Board, Sacramento, California.

Siegel, S., C. Toms, D. Gillenwater, C. Enright. 2010. Suisun Marsh tidal marsh and aquatic habitats conceptual model, chapter 3: tidal marsh. Suisun Marsh Habitat Management, Preservation, and Restoration Plan. http://www.fws.gov/sacramento/es/documents/Tidal_marsh_2010/Tidal_CM_Ch3_Tidal_Marsh.pdf.

Singer, M. B. 2007. The influence of major dams on hydrology through the drainage network of the Sacramento River basin, California. River Research and Applications 23: 55–72.

Suisun Ecological Workgroup. 2001. Final report to the State Water Resources Control Board. California Department of Water Resources, Sacramento.

U.S. Bureau of Reclamation, U.S. Fish and Wildlife Service, and California Department of Fish and Game. 2011. Suisun Marsh Habitat Management, Preservation, and Restoration Plan. Final environmental impact statement/environmental impact report. U.S. Bureau of Reclamation, Sacramento.

U.S. Fish and Wildlife Service. 2010. Draft recovery plan for tidal marsh ecosystems of northern and central California. U.S. Department of Interior, Fish and Wildlife Service, Sacramento. http://www.fws.gov/sacramento/ES/Recovery-Planning/Tidal-Marsh/es_recovery_tidal-marsh-recovery.htm.

U.S. Geological Survey. 2012. Magnitude 3.3 earthquake 3km south-southwest of Green

Valley, California, October 8, 2012. U.S. Geological Survey Earthquake Hazards Program. http://comcat.cr.usgs.gov/earthquakes/eventpage/nc71855125#summary.

Williams, G. P., and M. G. Wolman. 1984. Downstream effects of dams on alluvial rivers. U.S. Geological Survey Professional Paper No. 1286.

Wintzer, A. P., M. H. Meek, and P. B. Moyle. 2011. Life history and population dynamics of *Moerisia* sp., a non-native hydrozoan, in the upper San Francisco Estuary (U.S.A.). Estuarine and Coastal Shelf Science 94:48–55.

Wright, S. A., and D. H. Schoellhamer. 2005. Estimating sediment budgets at the interface between rivers and estuaries with application to the Sacramento–San Joaquin River Delta. Water Resources Research 41: W09428.

Zedler, J. B. 1983. Freshwater impacts in normally hypersaline marshes. Estuaries 6: 346–355.

9

Alternative Futures
for Suisun Marsh

Peter B. Moyle, Amber D. Manfree, Peggy L. Fiedler, and Teejay A. O'Rear

Projected changes in climate over the next 100+ years will have major impacts on California, especially on the San Francisco Estuary. These changes are already underway and are expected to accelerate (Lund et al. 2007; Moyle et al. 2012). Impacts on the Estuary will include inundation by rising sea level of many acres of mud flats and low-lying marsh, increasingly frequent failures of levees and dikes that protect natural and developed areas, and greater variation in environmental conditions as a result of larger floods and longer droughts. These changes will occur on a backdrop of growing human populations, with increased demand for fresh water and concomitant changes in land use, in a naturally variable environment. Overall, the intensively managed, highly engineered land-and-waterscape constructed in the Estuary over the past 160 years is likely to prove inadequate to meet the many expectations Californians have for this region. These expectations include preserving iconic native plants and animals such as salt marsh harvest mouse and delta smelt. Suisun Marsh is also increasingly looked upon by planners around the state as a major refuge for the native estuarine biota, a view not always shared by Suisun Marsh landowners and others deeply involved in managing and conserving the present Marsh.

This conflict in views is not surprising. Suisun Marsh is a biota-rich wetland upstream of the salty, tidal San Francisco Bay and downstream of the mostly fresh and semi-tidal Sacramento–San Joaquin Delta. It is one of the most productive and complex natural areas in the San Francisco Estuary because of its dynamic geomorphology and predominantly brackish conditions. The Marsh has been successfully managed for wildlife-based recreation for many decades, so in outward appearance it has not changed much compared with other wetland areas of the San Francisco Estuary (chapter 3). Appearances are deceiving, however, and, as previous chapters have explained, the Marsh has undergone more anthropogenic change through the decades than is generally appreciated. In fact, for its entire ca. 6,000-year history, the Marsh has been strongly influenced by human actions (chapter 2). The decades ahead will bring even more dramatic changes to the Marsh, primarily through sea-level rise, but also with more variable precipitation, reduced sediment delivery, continued land subsidence, increased water temperatures, and, most assuredly, earthquakes. Salinity can be expected to increase from its current range, the Marsh becoming saltier on average. Changes in salinity, combined with changes in water and air temperature, will result in conditions becoming more variable and much less predictable, with strong effects on plant and animal life cycles.

In this chapter, we present visions of alternative scenarios for the future of Suisun Marsh, up to about the year 2100, using the diverse information from previous chapters. These future alternatives are based on the idea that while the Marsh is going to change dramatically in response to global and regional change, we have choices as to how we, as a society, react to these changes. At one extreme, we can "go with the flow" and accept what happens with little investment, leaving local Marsh landowners and fringing urban areas to their own devices to cope with inevitable environmental problems (e.g., flooding). At the other extreme, we can invest huge amounts of public and private funding to literally keep the rising tides at bay, to keep the Marsh as it is today as much as possible. The actual public choice is likely to be an evolving response to change, with reactions to emergencies, such as a large earthquake, a major flood, or a rapid extinction trajectory of a protected species, driving management actions in the short term. An optimal response, of course, would be to *set clear goals* for what our society would like the future Marsh to be like, structurally and functionally, and then to *work with environmental change* to create such a Marsh.

Here, we first review the long history of Marsh change and then present a conceptual basis for future Marsh management, followed by reviews of likely changes to terrestrial and aquatic habitats. We synthesize this information to present four scenarios as alternative futures. The scenarios are "cartoons" because there are an infinite number of alternatives when the finer details are considered. We hope that readers will use our visions to stimulate their own thoughts regarding what the Marsh should and could be like in the future, even if they are contrary to what we envision.

IN SUISUN MARSH, CHANGE IS A CONSTANT

We cannot overemphasize that Suisun Marsh is not a static system (see figure 9.1), despite management viewpoints that generally treat it as if it were. In particular,

FIGURE 9.1. Suisun Marsh on January 31, 2011, after heavy rains and deliberate flooding by duck clubs. The salinity control gates on Montezuma Slough are visible in the lower left corner (photo by P. B. Moyle).

during the short period of Euro-American settlement, changes have been more pronounced than they had been for at least the preceding 6,000 years.

Climate over the past 4,000 years in the Sacramento and San Joaquin river watersheds included much more variability than we are accustomed to today. European explorers and colonists happened to arrive at the beginning of a relatively stable and benign period in California's climate history. Strong trends in paleoclimate records indicate that present residents can look forward to both massive floods (Dettinger and Ingram 2013) and droughts that will last a few years to hundreds of years (Malamud-Roam et al. 2007). This inherent regional variability will increase as a result of the more extreme conditions that climate change is expected to bring.

Sea levels have risen over the past 10,000 years, generally increasing slowly enough to allow marsh accretion in San Francisco Estuary lowlands to keep pace with water depth. However, sediment cores indicate that 8,000 to 10,000 years ago there was a period of more rapid sea-level rise, during which marshes on the periphery of the nascent San Francisco Bay drowned when waters rose faster than plants could build soil. The Suisun Marsh with which we are familiar today thus came into being over the past ca. 6,000 years in response to slow sea-level rise, as well as moderate precipitation patterns and ambient temperatures (chapter 2). During this time the former river valley of Suisun Valley Creek and its tributaries were inundated; sediment was deposited by rivers, tides, and wind; and new marsh habitat was colonized by water-loving plants that built a thick layer of peat soils. Organic matter and sediment accretion resulted in the fairly homogeneous topography of today's Marsh: low-lying estuarine marsh surfaces tend to be within a few decimeters of mean high tide levels, suggesting a dynamic equilibrium between deposition, erosion, and natural subsidence (Pestrong 1972). Levees along the tidal sloughs rise only a little higher.

However, today the sediment delivered by runoff that was crucial in building new marshland has been largely cut off by water infrastructure (Atwater et al. 1979; Malamud-Roam and Ingram 2004). In addition, vertical accretion of organic matter to create peat soils has historically matched rising water levels at a rate of about 20 cm/100 yr (Weir 1950; Schlemon and Begg 1975), although this rate may be low. Miller et al. (2008) experimentally demonstrated organic matter accretion in the Delta of, on average, 4 cm/yr, with maximum rates of 7–9 cm/yr. In short, wetlands of Suisun Marsh have generally had a benign relationship with sea-level rise in the past, but the accelerated rise underway is likely to make it difficult for Marsh vegetation and sediment deposition to keep up with the rising tides.

Brackish conditions in the Marsh that result from its intermediate position in the Estuary as well as its recent history create habitat for numerous important species of plants, migratory birds, mammals, and fishes. However, in the past there have been long periods of drought, causing more saline conditions (1,300–1,600, 800–1,000, and 200–300 years before present) as well as periods of above-average precipitation leading to fresher conditions (before 2,000 years ago, 1,200–1,300 years ago, and 200 years ago to present) (Malamud-Roam et al. 2007). Salinity in the Marsh and in the Delta has clearly been driven in part by regional precipitation patterns (figure 9.1), although today, water exports from the Delta result in an annual human-made drought for the Marsh (Malamud-Roam and Ingram 2004). To reverse this effect, salinity has been managed since the 1970s (Warner et al. 1971; Houghteling 1976). In Suisun Marsh, this is manifested in the salinity control gates located at the upper end of Montezuma Slough (chapter 3).

CONCEPTUAL BASIS FOR FUTURE MANAGEMENT

Suisun Marsh today is clearly a very different ecosystem than it was 150 years ago, despite abundant waterfowl and emergent marsh plants that create an illusion of constancy. Likewise, it will be a very different system 100 years from now. The Marsh is, and will be, a highly altered ecosystem containing many alien species that interact with the remaining native species. While there is much discussion about "restoration" of Suisun Marsh, there is little consensus on what condition(s) should characterize the future Marsh. Restoring it to the dynamic tidal marsh of the early 19th century is clearly not possible given the absence of the native Patwin peoples who lived there, the presence of abundant alien species, heavy development of land around the Marsh, and myriad other irreversible changes. Humans apparently have manipulated the biota of the Marsh since its beginning, and human influence is more evident now than ever. This means that we humans are responsible for the future of the Marsh and its biota, just as we have been in the past. We can actually direct that future, to a certain extent. Two interrelated concepts that seem particularly relevant to this directed future are reconciliation ecology and novel ecosystems.

Reconciliation ecology is defined as the "science of inventing, establishing, and maintaining new habitats to conserve species diversity in places where people live, work, and play" (Rosenzweig 2003: 7). Reconciliation ecology accepts the fact that humans increasingly dominate most ecosystems on the planet. Thus, to manage the Marsh, including restoration, managers should consciously determine what the ecosystem should look like and what species it should contain. If the Marsh were treated as a reconciled ecosystem in future planning, managers would presumably (but not necessarily) try to manage the system to favor a well-defined suite of desirable plants and animals, mostly natives, that require estuarine conditions (Moyle et al. 2012). The suite of interacting species should also provide ecosystem services such as hunting, recreation, water treatment, habitat for endangered species, and aesthetic views. The reconciled ecosystem might bear only modest resemblance to the 19th-century wetland ecosystem, because many current species of alien plants and animals will continue to be players in the future ecosystem.

The key to maintaining desirable aquatic species and conditions, of course, is active management toward a defined set of species and ecosystem goals. In the case of Suisun Marsh, goals could include maintaining wetlands that provide sweeping vistas of tules and cattails, support abundant wintering waterfowl, accept tertiary-treated wastewater for final polishing, and provide fishing for native and alien species. But drastically different goals could be set as well, as reflected in the four scenarios that follow.

Novel ecosystems are most readily recognized by being composed of a diversity of plants, invertebrates, and vertebrates from many parts of the world that have never coexisted before (Seastedt et al. 2008, Moyle 2013). The dominant players can be either native or alien species. The diverse species often seem to function together much like the coevolved members of a "natural" ecosystem. But novel ecosystems are often quite different from the original ecosystems they replaced. The difference is reflected in extinctions of native species, rapid additions of alien species (to greatly increase species richness), and presence of species and ecological processes that did not exist at the ecosystem location before. Because the assemblage of organisms making up a novel ecosystem is of recent origin, rapid change in ecosystem structure can be expected as species come and go or become abundant and then decline. Novel ecosystems are typically present in areas greatly altered by humans physically or chemically, through habitat modification or pollution.

Suisun Marsh today is a good example of a novel ecosystem. Alien species interact with native species at all levels, from plants to invertebrates to fish to mammals, in a landscape dominated by dredged sloughs and diked, heavily managed marsh "islands." Approximately half the fish species are aliens, originating from Germany, China, Japan, and eastern North America, but they form seemingly integrated food webs with alien and native fish and invertebrates. The dikes simultaneously provide habitat for alien weeds, like perennial pepperweed, which can provide refuge during high tides for endangered rails. What all this means from a management perspective is that physical habitat can, if managers desire, be manipulated in ways that favor desirable species, which are usually but not always native (e.g., striped bass). But it has to be recognized that all species interact with some other species, so management actions in a novel ecosystem may have unintended consequences. Thus, increases in salinity and tidal action in the Marsh, good for many native species, may facilitate invasion by alien overbite clam, a species whose filter-feeding diminishes food supplies for pelagic fishes such as native delta smelt and longfin smelt.

Taken together, these two concepts suggest that it is possible to take new and positive approaches to managing the Marsh. And the Marsh *does* need to be managed, and managed intensively. If Suisun Marsh is to continue to be an aesthetically pleasing ecosystem that is home to a diverse flora and fauna, especially to California's declining native biota, it will have to be managed creatively as well, recognizing that today's Marsh is an ecosystem that has never existed before. It is also an ecosystem of which we humans are irrevocably a part.

THE FUTURE OF MARSH (TERRESTRIAL) HABITATS

Rapidly rising sea level, combined with land subsidence and low sediment supplies, can be expected to cause at least seasonal inundation of much (perhaps all) of the present marshland, because even the highest areas of open marsh lie less than a meter above high tide. Various estimates bracket sea-level rise by year 2100 at 50 to 200 cm above recent levels (Rahmstorf 2007; Pfeffer et al. 2008; Nicholls et al. 2011), although other estimates are even higher. At a rise rate of less than 50 cm per century, it is possible that sections of the Marsh still connected to tidal, stream, and flood processes could accrete sufficient sediment and organic matter to match the rate of rising waters, but at 200 cm/100 yr, the likelihood of this happening is almost nonexistent. Following Knowles (2010), we assume a 100 cm increase in sea level by 2110; without management to allow accretion, this rise would convert much of the present Suisun Marsh to aquatic habitat with marshland habitat found mainly on the fringes, especially on the northeast end of the Marsh. At the same time, intertidal zones may shrink with decreased tidal range in the Delta as water spreads over larger areas and more and more highly subsided areas become flooded (chapter 3), making inundation constant in the lowest-lying parts of the Marsh. Higher storm surges may also affect the future geomorphology of the Marsh. One corollary of these saltier and wetter conditions will be alteration of the Marsh's plant communities, with the current freshwater and brackish-water species of the Marsh being succeeded by more salt-tolerant plants (chapter 4).

As the Marsh environment becomes increasingly variable in salinity, terrestrial vertebrate assemblages will be stressed. California's native fauna, having evolved under such conditions, may benefit unless, as is predicted, the new environment experiences greater extremes in conditions than can be accommodated even by native species. Drawing broad conclusions is difficult, however, because Marsh habitats are highly fragmented, areas of historical refuge no longer exist, and many alien species are abundant.

Changes in air and water temperatures also will play a role in determining ecological function. Plant and animal metabolisms have optimal ranges of temperature, and if the environment is too cold or too warm, certain species may be forced out of the area. Larger mammals and birds have considerable capacity to move to more favorable conditions. However, smaller, less mobile animals, such as Suisun ornate shrew and Suisun song sparrow, have more difficulty making such moves, especially at the rate at which climate is changing.

We can be fairly certain that the Marsh will keep its character as "open space" in the next century, but what that open space will look like depends on management actions we take today. In particular, its ability to support large populations of breeding and overwintering waterfowl will change. The present Marsh is a patchwork of 158 duck clubs, habitat restoration sites, and large tracts of public land managed for open space, hunting, and waterfowl habitat. The management goals for much of this land focus on providing habitat appropriate for freshwater-dependent waterfowl that breed or winter in the Marsh.

The view of local resource managers is that the future Marsh can be similar to the existing Marsh, with some additions of tidal habitat, during the 30-year time frame covered in the draft *Suisun Marsh Habitat Management, Preservation, and Restoration Plan* (hereafter "Suisun Marsh Plan"; U.S. Bureau of Reclamation et al. 2011):

> The preferred alternative . . . includes restoring 5,000 to 7,000 acres in the Marsh to fully functioning, self-sustaining tidal wetland and protecting and enhancing existing tidal wetland acreage; and improving the remaining 44,000 to 46,000 acres of managed wetlands, levee stability, and flood and drain capabilities.

In the wetland restoration sites, the Suisun Marsh Plan recognizes that breaching dikes for tidal restoration will initially create subtidal (open water) habitat. However,

> [T]he amount of subtidal aquatic habitat is expected to decrease gradually as sediment accretes and emergent tidal vegetation is established at each restoration site. As this happens, the site will be restored to a tidal wetland. However, the rate of accretion and the rate of sea level rise will dictate the end result, and the actual timeframe for such progression depends on the site-specific conditions, but significant geomorphic changes are decadal. Locations with large subsidence and low sediment concentrations may never return to emergent marsh and instead remain as open water. Adaptive management also will be used to improve restoration designs to achieve desired results.

Thus, the Suisun Marsh Plan seems to recognize that change can happen on a limited and manageable scale and that the rest of the Marsh can be "protected" by improved dikes and other actions, at least for 30 years. On this time scale, the above statements may indeed represent reality, barring a major earthquake or major flood. Unfortunately, on a longer time scale, larger changes to the entire Marsh seem inevitable (chapters 3 and 6). For example, many of the reinforced dikes may collapse simultaneously, as the forces of subsidence, sea-level rise, winds, and floods coincide. Some relief may come if a large-scale collapse of levees surrounding Delta islands occurs first, because the reduction in tidal energy caused by a huge body of open water in the Delta will reduce pressure on Suisun Marsh dikes. Sea level will keep rising, however, and the probability of a major earthquake will keep increasing.

Overall, the future of many terrestrial vertebrates and invertebrates in the Marsh will depend on a management strategy within the present Marsh that maintains large tracts of emergent marsh vegetation and on allowing marsh vegetation to expand into upland fringe areas that are not already urbanized.

THE FUTURE OF AQUATIC HABITATS

As with terrestrial habitats, the aquatic habitats of the future Marsh will be very different from those of today if present trends continue. However, exactly how these habitats will change is subject to great uncertainty, given the extreme complexity of forces affecting Marsh water quality and quantity. Nevertheless, numerous studies give us insight on the future of aquatic habitats in the Marsh (chapters 3 and 8).

Despite increasing efforts to control water that flows through the San Francisco Estuary and its watershed over the past 150 years, variation in precipitation associated with its Mediterranean climate has been acutely felt throughout the system. During most years, winter and springtime flows are successfully captured by dams and subsequently released during the summer, primarily for agricultural uses. Very high flows mostly pass through the reservoirs, resulting in widespread flooding on both engineered and natural floodplains (Enright and Culberson 2009). The combined effects of the State Water Project and the Central Valley Project have tended to reduce flows during springtime while elevating them in summer, with concomitant increasing and decreasing salinities in the Marsh, respectively (chapter 3). Nevertheless, high variability of precipitation since the State Water Project began operating is reflected by a correspondingly large fluctuation in Delta outflow and, consequently, more variable salinity in the Marsh. Because of the limited capacity of existing water-control infrastructure to contain these extreme effects of climate, it is very likely that effects of climate change on freshwater flows into the Marsh will be felt even more strongly in the future than they are today. This is given additional credence by the fact that the climate of California over the past 150 years has been a more stable and mild subset of a much more dynamic climate that has existed during the Holocene (chapter 2; Malamud-Roam et al. 2007).

Although there is still great uncertainty regarding many of the effects of climate change, one thing seems certain: the water will be warmer (chapter 3). While direct effects of warmer temperatures will doubtless have a significant impact on aquatic organisms of the Marsh (e.g., changing the timing and duration of spawning periods), the indirect effects of warmer climate on hydrology will be just as profound. In particular, warmer temperatures will likely lead to an increased rain-to-snow ratio and earlier melting of the Sierra Nevada snowpack, both of which will result in larger, earlier peak flows than are now common (chapter 3). Concurrently, the earlier runoff timing will result in a lower bank of water available for release during summer, causing reduced base flows during summer and autumn.

These hydrological changes, aside from changing the timing, magnitude, and duration of freshwater flows into the Marsh, will strongly affect the salinity regime of its channels and sloughs. In general, most years in the future will see a Suisun Marsh that is saltier than it is now, interspersed with years containing transient freshwater conditions during winter, followed by a return to saltier conditions during spring, summer, and autumn. These effects will be exacerbated by rising sea level and greater intrusion of marine water into the Marsh. This scenario does not even take into account naturally or human-caused changes to the Delta, such as flooding of central Delta islands or construction of a peripheral canal or tunnel, which could significantly alter tidal range and hydrology even further. Flooding of diked, subsided wetland areas will most likely create large expanses of open water dominated by aquatic macrophytes (chapter 4). Such major changes in aquatic habitats will clearly affect fish and invertebrate assemblages, favoring a different subset of species than are favored by present conditions.

Overall, climate change and rising sea level will result in Marsh aquatic habitats that will be saltier, although occasional large flows will freshen the Marsh during winter and early spring. Water in the Marsh also will be clearer and warmer, with temperatures during summer potentially reaching stressful levels for some species (e.g., adult striped bass). These changes will occur concurrently with a greater area of shallow open-water and channel habitat and an increase in the perimeter of aquatic edge habitat, which presumably will be good for juvenile fish. These changed environmental conditions will result in a fish community that has different pelagic fishes, more benthic fishes, and more native marine fishes (chapter 7).

ALTERNATIVE FUTURES

As the previous chapters show, major changes to the topography, physical processes, and ecology of Suisun Marsh are inevitable within the next 100 years, perhaps within the next 25 years. The major drivers of change will likely be sea-level rise, climate change (increased frequency of large floods and long droughts), earthquakes, land subsidence, and human alteration of the Marsh and surrounding areas, much as in the Delta (Lund et al. 2007, 2010). These changes will result in major changes to the plant and animal communities. But the nature of the future Suisun Marsh will depend on how change is managed or not managed. Here, we envision four alternative future states, recognizing that the scenarios represent extremes in possible conditions: *Fortress Marsh, Flooded Marsh, Reconciliation Marsh,* and *EcoMarsh.* However, in all four alternatives, Suisun Marsh remains as a key wetland ecosystem in the San Francisco Estuary, important habitat for native species, and open space for surrounding urban areas. It is a landscape that should provide protection for cities from sea-level rise, as well as growing recreational opportunities.

Fortress Marsh

Under this scenario (see map 13 in color insert), governments at federal, state, and local levels decide the Marsh must be protected in its present configuration at all costs, in order to

- maintain duck hunting at present levels,
- provide a buffer for urban areas from sea-level rise,
- protect native mammals and birds, especially endangered species,
- maintain examples of increasingly rare types of freshwater marsh ecosystems,

- maintain nursery areas for desirable fishes, especially native species, and
- maintain open space and recreational opportunities.

Protection at this scale requires a high level of new infrastructure and site manipulation, but it is essentially the goal of the Suisun Marsh Plan, which includes 7,000 acres of tidal marsh restoration (about 13% of the Marsh), but mostly imagines the status quo continuing.

Under this scenario, a Dutch-style massive dike would be built along the southern edge of the Marsh to protect it from major incursions of water from Suisun Bay. In addition, dikes along major sloughs would be reinforced to protect Grizzly Island, Joice Island, wetlands along Denverton and Suisun sloughs, and marshlands west of Suisun Slough. The railroad bed running across the western portion of the Marsh would be raised and reinforced, serving as an interior barrier to provide additional protection for approximately 25% of the Marsh and to capture freshwater inflows from Green Valley and Suisun creeks as well as the outflows of the Fairfield–Suisun Wastewater Treatment Plant. The major sloughs would remain open and tidal, if heavily restricted by dikes and tidal gates, although presumably Rush Ranch would retain more open tidelands. The salinity gates on Montezuma Slough would continue to operate, but time of operation would likely be extended into summer to capture as much of the inflowing water from the Sacramento River as possible. Additional gates would have to be built to protect interior areas from storm surges and major flood events.

Following the Suisun Marsh Plan, the Marsh would be managed as four distinct regions (map 13), presumably with their own set of rules for management. Each region would have acreage set aside and managed as restored tidal marsh, with connections to Montezuma or Suisun sloughs. With the combination of outside influences (chapter 3) and altered gate operation to maintain fresh water in the channels, tidal range in the restored marshes would be muted, with concomitant effects on vegetation, depending on location in the salinity gradient in the major sloughs.

Management of captive marshlands behind dikes would necessarily be intensive. Subsidence reversal using accretion of vegetation would be encouraged to keep up with sea-level rise, especially on Grizzly and Joice islands and in the western Marsh, reducing the impacts of inevitable dike breaches. Dredge spoils might be needed to augment sedimentation in targeted areas (Knowles 2010), as was done in the Montezuma Wetlands Project on the southeast edge of the Marsh (U.S. Army Corps of Engineers and Solano County 1998).

Even more than they are today, regions of the Marsh would be managed selectively for different species of protected, rare, and threatened animals and plants, such as splittail, delta smelt, salt marsh harvest mouse, California clapper rail, tule elk, and Suisun thistle. A special management emphasis would be on wintering and breeding waterfowl, perhaps including creation of more ponds, such as those on Joice Island. However, reduced connectivity (through the two slough entrances only) with other parts of the Estuary would be a problem for many species, especially migratory fish species, as well as for exchange of sediments and nutrients. With more stable conditions, it is likely that large populations of either overbite or Asian clams (or both) would develop in the sloughs, reducing food available to pelagic fishes. Indeed, it is likely that Fortress Marsh would a difficult place for pelagic species such as longfin smelt, delta smelt, and juvenile striped bass to live, as well as for juveniles of anadromous salmon and American shad.

The biggest problem we envision with this scenario, besides the exorbitant cost, is keeping up with physical pressure on the dikes created by the combination of sea-level rise, big floods, and occasional seismic events. It is easy to imagine a gradual abandonment of sections of the main dike and many interior diked areas as maintenance becomes increasingly difficult and expensive, unless there were strong political pressure (and money) to maintain the outer dike system to protect infrastructure on the edges of the Marsh, such as Suisun City, Fairfield, the railroad, and major highways.

Flooded Marsh

What if many of the dikes in Suisun Marsh were suddenly breached as the result of a major earthquake or of huge outflows from successive massive rainstorms, causing widespread flooding? If the breaches were too big and too numerous to fix, the great majority of the Marsh would be under water, at least at high tide. The later in the century this event happened, the deeper the newly flooded marsh would be (as shown in map 14 in color insert), assuming (reasonably) that no major changes to Marsh infrastructure were made. Under this scenario, Marsh landowners and agencies would be left to cope with flooding on their own, with little additional help from local, state, or federal governments. The state mandate to protect the Marsh with freshwater inflows and tidal gates would be considered impossible or uneconomical to accomplish and consequently lifted. While some dikes would no doubt be repaired, especially in the western Marsh, marshlands would be largely left to endure sea-level rise without further human intervention.

With the breaching of the highly vulnerable outer dikes on Suisun, Honker, and Grizzly bays, the present freshwater distribution system, especially that on Grizzly Island (Roaring River Slough), would be lost, resulting in abandonment of the tidal gates on Montezuma Slough. Depending on the amount and rate of flooding caused by sea-level rise, much of the Marsh would transform into tidal or subtidal habitat, with some of the latter becoming open water, and shallower areas becoming filled with tules and other rooted aquatic plants. Former dikes would become linear islands until eroded away, while dendritic tidal channels would develop in some areas. Presumably, subsidence processes would be replaced by deposition processes in areas close to inflowing water from the Sacramento River, although suspended sediment supplies are limited (Schoellhamer et al. 2005). Where possible, subsidence reversal using deposition of plant matter from tules and other marsh plants would be encouraged by special management, perhaps funded in part to support carbon sequestration. If tule and other plant growth could keep pace with sea-level rise, diverse tidal marsh habitats would be created, including deep ponds surrounded by emergent plants that would likely support beds of submerged aquatic vegetation (chapter 4). However, maintaining emergent vegetation is problematic if sea-level rise exceeds 2–3 cm/yr.

The amount of high tidal habitat, important for harvest mice, clapper rails, and other marsh-dependent vertebrate species, as well as the amount of transitional upland habitat for native plant species, would depend on the degree to which the land had subsided before this large-scale flooding and how much land remained on higher-elevation contours after urban encroachment. On public lands, islands of higher elevation could be created as part of a conservation strategy to provide high-tide refuges.

The Southern Pacific railroad bed through the western Marsh would have to be abandoned, rerouted (e.g., along Highway 680), or greatly reinforced—and most surely raised. Maintaining the present route, which in our view is the most likely scenario, would essentially result in diking off the western Marsh from the rest of the Marsh unless larger bridges, with tidal flap gates, were created to accommodate increased tidal and creek-flood flows.

Under this scenario, Suisun Marsh would become a mixture of tidal, subtidal, and open-water habitat, with historical channels maintained in some places, in part by dredging. Salinities would likely show high variability, both seasonally and among years, depending on Sacramento River inflows and the degree to which flooding of Delta islands upstream reduced tidal action. The main channels of the Marsh would become wider and shallower, with many connections to interior marshlands, which today are mostly not directly connected. Such complex aquatic–marshland habitat would likely favor estuarine fish species such as striped bass, longfin smelt, splittail, and perhaps juvenile Chinook salmon. Productivity of invertebrates, favored as food by fish, could be comparatively high if variability in salinity, turbidity, and other variables were sufficient to discourage development of large populations of freshwater or brackish-water clams in the channels.

Most duck clubs would eventually cease to exist under the *Flooded Marsh* scenario, except possibly in the western Marsh, where the railroad dike would protect some of the less subsided lands. The new brackish habitat would presumably be suboptimal for most wintering waterfowl, although how much they actually used the Marsh each year would depend on the status of their habitat wildlife areas and rice fields throughout the Delta and the Sacramento and San Joaquin valleys. In exceptionally dry years, the Marsh could still be a refuge for waterfowl, especially if pond areas supported submerged aquatic vegetation suitable for use as food. *Flooded Marsh* could be an effective barrier against wave action from sea-level rise for urban areas, if vegetation growth kept pace with the rise and there were strategic placement of low levees, combined with abandonment of urbanized fringe areas subject to chronic flooding.

Reconciliation Marsh

The draft Bay Delta Conservation Plan requires large-scale restoration of tidal marsh, perhaps around 25,000 acres in the Delta and vicinity, to mitigate for the effects of proposed new water diversion infrastructure. One of the places often discussed for such mitigation habitat is Suisun Marsh. In essence, the *Reconciliation Marsh* scenario (see map 15 in color insert) assumes that much of the Marsh will be managed as tidal marsh to compensate for wetland habitats lost elsewhere. This means that large sums of public money would be available to carefully accommodate sea-level rise, including land purchase, while providing diverse habitats for desirable species and satisfying public demands for other traditional Marsh uses such as hunting.

This scenario includes many aspects of the other scenarios. Some of the "reconciliation" could include the following generalized actions (in no particular order):

· Creation of accommodation space for sea-level rise by preparing current higher-elevation marsh areas to flood and by allowing fringe areas to become available for some flooding, such as the area around upper Denverton Slough.
· Creation of fortress-like dikes around Grizzly and Joice islands, their interiors intensively managed habitat for waterfowl, including summer reproduction, as well as for various other native species such as rails. These islands would also be used for carbon sequestration (also a potential source of funds) and subsidence reversal.
· Provision of limited or no protection for islands and tidelands along Suisun Bay on the assumption that such lands, even if not behind levees, can naturally accrete sediment from Sacramento River sources, possibly keeping up with sea-level rise. If they can't keep up, they will become submerged baylands.
· Management of the region west of the railroad mainly as duck clubs, using freshwater input from the wastewater treatment plant and inflowing creeks and sloughs, when outflows and tides are right. A secondary goal for this region is subsidence reversal through organic matter deposition. The railroad bed would presumably be raised and reinforced to protect against wave action and subsidence, while flap gates or tidal gates would be installed where sloughs flow under railroad bridges.
· Management of duck clubs and wildlife areas in ways aimed at improving

future conditions in the Marsh in general. Actions would include reversing subsidence, improving water quality in surrounding channels (e.g., limiting black-water discharges), creating dikes that have increased habitat functionality (e.g., more gradual back slopes with extensive native vegetation, including plants characteristic of wetland–upland ecotones), providing native plants as food crops for ducks, and providing more habitat for endangered rails and other animals and plants.

- Creation of wildlife corridors that connect Grizzly Island and the Potrero Hills to Jepson Prairie and similar grassland landscapes. Tule elk would graze the hills and have free access to other areas as far away as the Yolo Bypass.
- Removal or setting back of dikes along some channels to create more fringing marsh as fish habitat, as well as improving natural connections to interior regions where possible (i.e., reconnecting sloughs). This would be especially relevant in the Nurse–Denverton Slough area, where such edge habitat already exists in many areas and some interior connections have been established (e.g., Blacklock Island), but set-backs could be applied wherever opportunity exists in the Marsh.
- Provision of protection for urban areas from flooding from sea-level rise and climate change, provided that these areas do not expand farther into the present Marsh. Ideally, such dikes would be inland far enough to allow expansion of tidal marsh habitat.
- Management of the Denverton–Nurse slough area as key habitat for rearing native fishes, such as splittail and delta smelt, by finding ways to enhance its productivity and attractiveness for fish spawning and rearing.
- Planning for conditions that will result from freshwater inflow being increasingly less available as climate change produces extended droughts and as large-scale changes to the Delta (e.g., island flooding; Moyle 2008) capture more of the river inflow. This will mean expecting long periods of fairly salty conditions that will favor pickleweed and similar plants, followed by periods of flooding with fresh water. It will also become necessary to manage some parts of the Marsh as refuges for brackish-water species during long periods of drought.
- Determination of whether the tidal gates on Montezuma Slough are still needed to maintain the diverse biota of the Marsh, including waterfowl. Consideration should also be given to managing the gates differently. For example, they could be used experimentally to keep Montezuma Slough brackish, rather than fresh, to create habitat for rearing delta smelt in summer and fall. This would require reversing the way the gates are operated today.

This third scenario basically sees the Marsh as a fairly independent entity that would retain waterfowl rearing and overwintering as important ecosystem functions, although greatly reduced from present levels. Management would have to be adaptive to realistically accommodate sea-level rise, as well as the needs of the diverse native plants and wildlife. One of the main functions of *Reconciliation Marsh* would be to provide wildlife habitat that aids in the recovery of threatened or endangered species, as required by the Bay Delta Conservation Plan.

EcoMarsh

The goal of the *EcoMarsh* would be to maximize habitat diversity for native animals and plants in the Marsh, while also managing for landscape connectivity to nearby terrestrial and aquatic habitats. At the same time, the *Ecomarsh* would accommodate sea-level rise and other changes in ways that help to protect urban areas around the Marsh and enhance provision of ecosystem services such as sewage disposal and recreation. In this scenario, managers would work with the forces of change, rather than fighting them, to transform the Marsh into a highly managed novel ecosystem.

The transformation would start with a comprehensive vision for the *EcoMarsh* to allow for careful planning and adaptive marsh reconstruction over the years (see map 16 in color insert). This scenario includes major infrastructure changes, including the following:

- relocating the railroad to the western edge of the Marsh, allowing the entire Marsh to be managed as a single unit,
- relocating Grizzly Island Road to a higher elevation in the hills to provide for more and better marsh–upland transition habitat (ecotone),
- putting several miles of Highway 12 on causeway to allow for marsh expansion to the north,
- relocating urban boundaries (e.g., southern edge of Suisun City) to provide some room for marsh expansion and placement of protective levees,
- converting the Lawler Ranch Ditch back to a functioning wetland ecosystem, and
- actively manipulating higher-elevation areas on the marsh plain to create tidal habitats that provide productive rearing areas for different life stages of diverse fishes, while also providing habitat for waterfowl and other native wildlife (e.g., as in Luco Slough today).

None of these actions can occur quickly. Thus, while natural resource planning and adaptive experimentation were taking place, much of the Marsh would be managed for subsidence reversal, carbon sequestration, and pond creation. Ponds would be restored to create conditions that once existed in much of the western Marsh before land conversion. This restoration would require maintaining most of the dikes for 20–30 years, until the interiors could be flooded and still support dense growths of marsh vegetation.

An important part of this scenario is managing the Marsh as an interconnected system of habitats. From a fish's perspective, the Marsh would be the bottom end of an arc of habitat that starts with the Yolo Bypass, extends through the expansive Cache–Lindsey slough region (potentially another major wetland restoration region), continuing down the Sacramento River, through Sherman Lake, and then on to the Marsh. From a wildlife perspective, an important movement corridor would connect Grizzly Island and the Potrero Hills with the native grasslands and vernal pools of Jepson Prairie Reserve. Other wildlife corridors could also be developed, if less easily, along Green Valley and other inflowing creeks along the western portion of the Marsh. As part of this landscape connectivity, some nearby areas outside the Marsh, such as Sherman Island, could be taken out of agriculture (mostly livestock pasture) and converted into waterfowl hunting areas.

In many respects, the *EcoMarsh* is a combination of *Flooded Marsh* and *Reconciliation Marsh* because large areas become tidal and subtidal habitat to accommodate sea-level rise. Most of the numbered points covered in the *Reconciliation Marsh* section also apply to the *EcoMarsh*. However, the *EcoMarsh* scenario assumes bold, proactive management that follows a long-term visionary plan, using adaptive management in the way it is supposed to be used (i.e., as experimental, informative, iterative land management). It also assumes, of course, that most Marsh property owners would cooperate with it, in part because the first 30 years do not require deviation in many ways from the *Suisun Marsh Plan,* as long as duck clubs are willing to change management practices to reverse subsidence and not degrade water quality. However, an *EcoMarsh* would also be a center of experimentation in Marsh management. For example, fire could be used as a management tool in some areas to recreate habitat similar to what the Patwin people created with their presumed regular burning of tule marshes.

In the longer run, management options could include sculpting of existing marshlands, adding sediment, removing selected dikes, and creating raised areas with the dike material. In addition, management would maintain some existing large channels through dike maintenance, but with dikes designed to overtop or with gates to handle high inflows. The higher-elevation or edge parts of the Marsh, such as Peytonia Slough Preserve, could be diked and managed specifically for threatened species such as salt marsh harvest mouse. All undeveloped areas fringing the Marsh, such as the Denverton Creek watershed and the Montezuma Wetlands restoration site, would expand to tidal marsh, integrated into the entire Marsh system. Potrero Hills, including the site of the present-day landfill operation, would therefore be managed as upland habitat for tule elk, grassland birds, and similar animals, as well as for native plants.

Caveat

For all four scenarios presented above, an underlying assumption is that within a hundred years we humans will have gained some control over climate change and sea-level rise. If the planet continues to warm, then eventually marsh habitat will try to expand into places where urban areas are today. Massive dikes, Dutch style, will be required to protect cities if such is indeed possible. Suisun Marsh will become an extension of Suisun Bay. The former Marsh will be sublime for boating and wind surfing, but not for most of the other things the public values it for today.

CONCLUSIONS

Suisun Marsh is a highly altered ecological sanctuary in a human-dominated landscape. It is also a dynamic place at many different scales of time, from daily to decadal. In the past 150 years or so, it has evolved into a region of fragmented land management that nevertheless favors many desirable species, including rare plants, native fishes, diverse waterfowl, and tule elk. These species are imbedded in a novel ecosystem that contains many alien species. The Marsh is now facing a period of accelerated irreversible change brought on by human activity, especially global climate change. Sea level and temperatures are already rising, formerly extreme conditions are less so, and predicted big floods and long droughts seem increasingly likely. At the same time, California's citizenry is increasingly dependent on water diverted through the Delta, and reduction of freshwater inflow is increasingly a factor in Marsh ecology and management. Most of the Delta is so severely altered that planners are looking elsewhere for succor for endangered species and habitats. Suisun Marsh, Cache–Lindsey Slough, and a few additional areas on the fringes of the Delta quickly rise to the top of the list of places to "save" for their natural values.

However, even these favored areas are highly altered, and their future depends on intensive management, with a strong human presence. We Californians have to decide what we want the Marsh to be like in 50–100 years and what species we want to have with us in that time. Ideally, we would find ways to reconcile human demands on the future Marsh with societal goals such as maintaining assemblages of native species and preventing extinctions. But this is unlikely to happen in a systematic fashion, given that we mostly manage resources in response to crises. Nevertheless, we should at least have a general idea of what future we want for Suisun Marsh so that our responses to crises can be deliberate rather than the result of panic.

We have presented four scenarios, representing quite different futures, to provide some general ideas as to alternative futures for the Marsh based on actions we take in the next few decades. The good thing is that the Marsh will remain "open space" no matter what we do and there will always be some kind of wetland, or at least wet, ecosystem there. But the functioning of this open space from the perspective of the biota and ecosystem services it can support will vary immensely depending on what actions we take.

The four scenarios can be looked at as possible outcomes of a series of large-scale experiments in landscape management, based on manipulating ecosystem processes (Euliss et al. 2008). Ideally these experiments would result from an adaptive management approach, whereby we learn from our successes and failures and expect to be surprised by results on occasion because of our imperfect knowledge of ecological systems. For example, it is easy to imagine a study comparing different regions of the Marsh with different diking strategies (e.g., fortress-like protection vs. breached dikes) in terms of deposition rates of organic matter and carbon sequestration. Or to compare the fish assemblages and invertebrate production of different sloughs in areas with different management strategies. Or to compare different grazing management strategies for the Potrero Hills, including comparing livestock to tule elk as principal grazers. Or to systematically compare different vegetation "treatment" strategies on duck clubs to determine factors that affect duck production. The results of such comparisons could be used to guide management decisions for the future and provide new strategies for dealing with the effects of sea-level rise and climate change. To make an adaptive management program a reality, however, it has to have the following characteristics:

- A clear set of widely agreed-upon goals and objectives for the program. These should include social, economic, and political goals and objectives, as well as biological ones.
- Monitoring of key organisms (e.g., fish, wading birds, ducks, marsh plants) on at least an annual basis, along with continuous monitoring of environmental variables such as water quality and temperature.
- A centralized, publically accessible system for data gathering and deposition.
- A multidisciplinary team of scientists willing to design management experiments, guide monitoring, analyze data, and publish results.
- Strong involvement by local stakeholders.
- Strong persistent leadership.

Such a program of reconciliation ecology would require significant funding every year, but it might become self-sustaining by reducing the costs of responding to emergency dike repair, fish and waterfowl kills, and similar problems, and through payments for carbon sequestration.

We, the editors of this book, believe that creating a reconciled Marsh using adaptive management strategies is both possible and desirable. We envision a Marsh with many of its present ecosystem functions increased (e.g., large expanses of tidal marsh teeming with birds and fish, dominated by native plants). We also envision a Marsh as a major node in a network of natural areas. Such a Marsh will be a magnet for people who wish to experience a tiny bit of wildness in an urbanized landscape, a balm for the "nature deficit disorder" (Louv 2005) that afflicts so many of us. But such a Marsh will not happen spontaneously. It will take vision, planning, and money. The sooner we start on this journey the better, because the accelerating effects of global climate change are reducing the number of choices we have for action.

REFERENCES

Atwater, B. F., S. G. Conard, J. N. Dowden, C. W. Hedel, R. L. MacDonald, and W. Savage. 1979. History, landforms, and vegetation of the Estuary's tidal marshes. Pages 347–385 in San Francisco Bay: The Urbanized Estuary (T. J. Conomos, A. E. Leviton, and M. Berson, eds.). San Francisco: AAAS Pacific Division.

Dettinger, M. D., and D. L. Ingram. 2013. The coming megafloods. Scientific American 398(1): 64–71.

Enright, C., and S. D. Culberson 2009. Salinity trends, variability, and control in the northern reach of the San Francisco Estuary. San Francisco Estuary and Watershed Science 7(2): article 3.

Euliss, N. H., L. M. Smith, D. A. Wilcox, and B. A. Browne 2008. Linking ecosystem processes with wetland management goals: charting a course for a sustainable future. Wetlands 28: 553–562.

Houghteling, J. C. 1976. Suisun Marsh Protection Plan. San Francisco Bay Conservation and Development Commission, San Francisco.

Knowles, N. 2010. Potential inundation due to rising sea levels in the San Francisco Bay region. San Francisco Estuary and Watershed Science 8(1).

Knuuti, K. 2001. Tidal wetland restoration: accelerating sedimentation and site evolution in restoring diked baylands. Seventh Federal Interagency Sedimentation Conference, Reno, Nevada.

Louv, R. 2005. Last Child in the Woods: Saving Our Children from Nature-deficit Disorder. New York: Algonquin.

Lund, J., E. Hanak, W. Fleenor, W. Bennett, R. Howitt, J. Mount, and P. B. Moyle. 2010. Comparing Futures for the Sacramento–San Joaquin Delta. Berkeley: University of California Press.

Lund, J., E. Hanak., W. Fleenor, W., R. Howitt, J. Mount, and P. Moyle. 2007. Envisioning Futures for the Sacramento–San Joaquin Delta. San Francisco: Public Policy Institute of California.

Malamud-Roam, F., D. Dettinger, B. L. Ingram, M. K. Hughes, and J. L. Florsheim. 2007.

Holocene climates and connections between the San Francisco Bay Estuary and its watershed: a review. San Francisco Estuary and Watershed Science 5(1): article 3.

Malamud-Roam, F., and B. L. Ingram. 2004. Late Holocene $\delta^{13}C$ and pollen records of paleosalinity from tidal marshes in the San Francisco Bay estuary, California. Quaternary Research 62: 134–145.

Miller, R. L., M. S. Fram, R. Fujii, and G. Wheeler. 2008. Subsidence reversal in a re-established wetland in the Sacramento–San Joaquin Delta, California, USA. San Francisco Estuary and Watershed Science 6(3): article 1. http://repositories.cdlib.org/jmie/sfews/vol6/iss3/art1.

Moyle, P. B. 2008. The future of fish in response to large-scale change in the San Francisco Estuary, California. Pages 357–374 in Mitigating Impacts of Natural Hazards on Fishery Ecosystems (K. D. McLaughlin, ed.). Symposium 64. Bethesda, Maryland: American Fishery Society.

Moyle, P.B. 2013. Novel aquatic ecosystems: the new reality for streams in California and other Mediterranean climate regions. River Research and Applications. DOI: 10.1002/rra.2709.

Moyle, P. B., W. Bennett, J. Durand, W. Fleenor, B. Gray, E. Hanak, J. Lund, and J. Mount. 2012. Where the Wild Things Aren't: Making the Delta a Better Place for Native Species. San Francisco: Public Policy Institute of California.

Nicholls, R. J., N. Marinova, J. A. Lowe, S. Brown, P. Vellinga, D. de Gusmao, J. Hinkel, and R. S. J. Tol. 2011. Sea-level rise and its possible impacts given a 'beyond 4°C world' in the twenty-first century. Philosophical Transactions of the Royal Society 369: 161–181.

Pestrong, R. 1972. San Francisco Bay tidelands. California Geology 25: 27–40.

Pfeffer, W. T., J. T. Harper, and S. O'Neel. 2008. Kinematic constraints on glacier contributions to 21st-century sea-level rise. Science 321: 1340–1343.

Rahmstorf, S. 2007. A semi-empirical approach to projecting future sea-level rise. Science 315: 368–370.

Rosenzweig, M. L. 2003. Win-Win Ecology: How the Earth's Species Can Survive in the Midst of Human Enterprise. Oxford: Oxford University Press.

Schlemon, R. J., and E. L. Begg. 1975. Late Quaternary evolution of the Sacramento–San Joaquin Delta, California. Pages 259–266 in Quaternary Studies (R. P. Suggate and M. M. Cresswell, eds.). Wellington: Royal Society of New Zealand.

Schoellhamer, D. H., M. A. Lionberger, B. E. Jaffe, N. K. Ganju, S. A. Wright, and G. G. Shellenbarger. 2005. Bay sediment budget: Sediment accounting 101. Pages 58–63 in Pulse of the Estuary: Monitoring and Managing Water Quality in the San Francisco Estuary. San Francisco Estuary Institute, Oakland, California.

Seastedt, T. R, R. J. Hobbs, and K. N. Suding. 2008. Management of novel ecosystems: are novel approaches required? Frontiers in Ecology and Evolution 6: 547–553.

U.S. Army Corps of Engineers and Solano County. 1998. Montezuma Wetlands Project, Final Environmental Impact Report/Environmental Impact Statement. State Clearinghouse (SCH) No. 91113031. U.S. Army Corps of Engineers, San Francisco, and Solano County Department of Environmental Management, Fairfield, California.

U.S. Bureau of Reclamation, U.S. Fish and Wildlife Service, and California Department of Fish and Game. 2011. Suisun Marsh habitat management, preservation, and restoration plan. Final Environmental Impact Statement/Environmental Impact Report. U.S. Bureau of Reclamation, Sacramento.

Warner, G. H., H. K. Chadwick, et al. 1971. Report to the State Water Resources Control Board on the impact of water development on the fish and wildlife resources of the Sacramento–San Joaquin Delta and Suisun Marsh. California Department of Fish and Game, Sacramento.

Weir, W. W. 1950. Subsidence of peat lands of the San Joaquin–Sacramento Delta, California. Hilgardia 20: 37–56.

GEOSPATIAL DATA SOURCES

CalAtlas. 2012. California Geospatial Clearinghouse. State of California. Available: http:// atlas.ca.gov/. Accessed: 2012.

California Department of Water Resources Suisun Ecological Workgroup. 2007. LIDAR dataset. Available by request. Accessed: June 2012.

Contra Costa County. 2013. Contra Costa County Mapping Information Center. Available: http://www.ccmap.us/. Accessed: January 2013.

Gesch, D., M. Oimoen, S. Greenlee, C. Nelson, M. Steuck, and D. Tyler. 2002. The National Elevation Dataset. Photogrammetric Engineering and Remote Sensing 68 (1): 5–11.

San Francisco Bay Conservation and Development Commission. 2010. Suisun Marsh Protection Plan. State of California. Available: http://www.bcdc.ca.gov/laws_plans/ plans/suisun_marsh#6. Accessed: August 2012.

Solano County. 2013. Geographic Information Systems Homepage. Solano County Department of Information Technology. Available: http://www.co.solano.ca.us/depts/ doit/gis/. Accessed: November 2012.

INDEX

Ackerman, Joshua T., vii, 103
adaptive management, 4, 89, 134, 203–04, 216,
 224, 225, 227, 228
agriculture, 2, 19, 21, 22, 31, 32, 33, 73, 84, 89, 103,
 108, 127, 139, 140, 141, 149, 151, 156, 195, 217
alien: species, 4, 22, 134, 141, 143–45, 155, 156–57,
 190–91, 212–15, 226; animals, 86, 135–38, 140,
 148, 152, 165–67, 169, 170, 172–73, 176–78,
 179–80; invasive, 90, 185; plants, 29, 69, 70,
 73–78, 82, 86, 88, 91, 120, 147. *See also specific
 species*
alternative futures, 218–226
American beaver. *See* beaver
amphibians, 2, 133, 138, 143, 150, 154–55
anadromous fish, 220
Antioch, 24, 25
Anza, Juan Bautista de, 30, 73
Ayala, Don Juan Manuel de, 26, 30

bass. *See* largemouth bass, striped bass
Bay Delta Conservation Plan (BDCP), 186, 192,
 197, 201
Baye, Peter R., vii, 65
bear, 25, 29, 32; black, 139, 152; grizzly, 27, 28,
 139, 152
beaver, 27, 29, 31, 38, 135, 139, 152, 153–54
Benicia, 3, 24, 28, 35
biodiversity, x, 22, 92, 133, 159
birds, 2, 25, 32, 49, 109, 122, 136–37, 139, 142; and

alien plants, 145; and disease, 146; migratory, 38, 92, 139, 140–41, 159; and pollution,
 146; predation on, 144–45, 152, 153; status
 and trends, 146–51
bird's-beak, soft, 67, 69, 76, 77, 84
black bear, 139, 152
Boynton Slough, 48, 54, 195
brackish, 12, 19, 22, 46, 47*fig.*, 66, 72–73, 81, 82,
 85, 88, 90, 93, 119, 120, 123, 125–26, 155, 212,
 221
Bradmoor Island, 21
Brazilian waterweed, 77
Brown's Island, 2, 77, 193
bulrush, 24, 29, 51, 68, 69, 70, 72, 73, 75, 80, 82,
 83, 84, 88, 120, 123, 127, 128
burning. *See* fire
Burns, Edward, vii, 103
butterflies, 2, 76, 85, 156–57

Cache Slough, 17, 225, 226
CALFED, 36, 192
California, 15, 24, 188; butterflies in, 156–57;
 climate of, 15, 209, 211, 217; Endangered Spe-
 cies Act, 200; history of, 31–32; waterfowl
 in, 104, 105, 106–9, 110, 104–19, 121–22, 127,
 128, 141; water rights, 202
California bay shrimp, 169, 170
California black rail, 141, 143, 147, 159
California bulrush, 24, 68, 80